Report of Investigations 9677

Key Design Factors of Enclosed Cab Dust Filtration Systems

By John A. Organiscak and Andrew B. Cecala

DEPARTMENT OF HEALTH AND HUMAN SERVICES
Centers for Disease Control and Prevention
National Institute for Occupational Safety and Health
Pittsburgh Research Laboratory
Pittsburgh, PA

November 2008

This document is in the public domain and may be freely copied or reprinted.

Disclaimer

Mention of any company or product does not constitute endorsement by the National Institute for Occupational Safety and Health (NIOSH). In addition, citations to Web sites external to NIOSH do not constitute NIOSH endorsement of the sponsoring organizations or their programs or products. Furthermore, NIOSH is not responsible for the content of these Web sites. All Web addresses referenced in this document were accessible as of the publication date.

Ordering Information

To receive documents or other information about occupational safety and health topics, contact NIOSH at

>Telephone: **1–800–CDC–INFO** (1–800–232–4636)
>TTY: 1–888–232–6348
>e-mail: cdcinfo@cdc.gov
>
>or visit the NIOSH Web site at **www.cdc.gov/niosh**.

For a monthly update on news at NIOSH, subscribe to NIOSH *eNews* by visiting **www.cdc.gov/niosh/eNews**.

DHHS (NIOSH) Publication No. 2009–103

November 2008

SAFER • HEALTHIER • PEOPLE™

CONTENTS

Page

Abstract 1
Introduction 2
Test apparatus and measurement methods 3
Experimental cab test factors 6
Experimental design 8
Experimental results 9
Statistical analysis 14
Mathematical model for cab penetration 15
Discussion 18
Conclusions 21
Acknowledgment 22
References 22
Appendix A.—Half-fraction experimental design 24
Appendix B.—Experimental test data 25
Appendix C.—Stepwise regression analysis of filtration system without pressurizer 34
Appendix D.—Stepwise regression analysis of filtration system with and without pressurizer 37
Appendix E.—Mathematical model for cab filtration system 40

ILLUSTRATIONS

1. Experimental cab test apparatus 3
2. Air pressure-quantity characteristic curves of fans on experimental cab apparatus 4
3. Laboratory cab test apparatus used in PRL's longwall test gallery 5
4. Box-and-whisker plot of cab *Pen* for filter combinations without pressurizer 9
5. Box-and-whisker plot of cab *Pen* for filter combinations with pressurizer 10
6. Relationship between intake filter differential pressure and intake airflow 12
7. Relationship between cab differential pressure and intake airflow 13
8. Relationship between intake leakage and intake filter differential pressure 13
9. Mathematically modeled cab *Pen* versus experimentally measured cab *Pen* 18
10. Ambient air size classified particle count concentrations 19
C-1. Standardized predicted values for regression model without pressurizer 35
C-2. Normal probability plot of standardized residuals without pressurizer 35
C-3. Standardized residuals versus standardized predicted values without pressurizer 36
D-1. Standardized predicted values for regression model with and without pressurizer ...38
D-2. Normal probability plot of standardized residuals with and without pressurizer 38
D-3. Standardized residuals versus standardized predicted values with and without pressurizer ...39
E-1. Schematic of basic cab filtration system 41

TABLES

1. Experimental cab test factors 6
2. Cab testing results without pressurizer 11
3. Cab testing results with pressurizer 11
4. List of statistically significant regression factors affecting cab *Pen* 14

CONTENTS—Continued

Page

   5. Recirculation filter efficiency results for 0.3- to 1.0-μm-sized particles 16
   6. Summary of NIOSH enclosed cab field studies 20
A-1. Half-fraction experimental design 24
B-1. First-half fraction test data without intake pressurizer ($I = +ABCDE$) 25
B-2. Second-half fraction test data without intake pressurizer ($I = -ABCDE$) 28
B-3. Pressurizer test data 31
C-1. Stepwise regression model without pressurizer 34
C-2. ANOVA for stepwise regression model without pressurizer 34
D-1. Stepwise regression model with and without pressurizer 37
D-2. ANOVA for stepwise regression model with and without pressurizer 37

ACRONYMS AND ABBREVIATIONS USED IN THIS REPORT

AAF	American Air Filter
ANOVA	analysis of variance
ASHRAE	American Society of Heating, Refrigerating and Air-Conditioning Engineers, Inc.
CFR	Code of Federal Regulations
HVAC	heating, ventilation, and air conditioning
MERV	minimum efficiency reporting value
MSHA	Mine Safety and Health Administration
NIOSH	National Institute for Occupational Safety and Health
PRL	Pittsburgh Research Laboratory (NIOSH)
PVC	polyvinyl chloride
STP	standard temperature and pressure

UNIT OF MEASURE ABBREVIATIONS USED IN THIS REPORT

ft	foot
ft^2	square foot
ft^3	cubic foot
ft/min	foot per minute
ft^3/min	cubic foot per minute
hr	hour
in	inch
in^2	square inch
in Hg	inches of mercury
in w.g.	inches of water gauge
L	liter
L/min	liter per minute
mA	milliampere
min	minute
mg/m^3	milligram per cubic meter
mph	miles per hour
V	volt
V dc	volt, direct current
μm	micrometer
°F	degree Fahrenheit

KEY DESIGN FACTORS OF ENCLOSED CAB DUST FILTRATION SYSTEMS

By John A. Organiscak[1] and Andrew B. Cecala[1]

ABSTRACT

Enclosed cabs are a primary means of reducing equipment operators' silica dust exposure at surface mines. The National Institute for Occupational Safety and Health experimentally investigated various factor effects on cab air filtration system performance. The factors investigated were intake filter efficiency, intake air leakage, intake filter loading (filter flow resistance), recirculation filter use, and wind effects on cab particulate penetration. Adding an intake pressurizer fan to the filtration system was also investigated.

Results indicate that intake filter efficiency and recirculation filter use were the two most influential factors on cab penetration performance. Use of the recirculation filter reduced cab penetration by usually an order of magnitude over the intake air filter alone because of the multiplicative filtration of the cab interior air. Intake air leakage and filter loading affected the cab penetration to a lesser extent, while wind had the least impact on cab penetration between the calm and 10-mph wind velocities tested. Adding an intake pressurizer fan notably increased intake airflow and cab pressure with only minor changes to cab penetration. A mathematical model was developed that describes cab penetration in terms of intake filter efficiency, intake air quantity, intake air leakage, recirculation filter efficiency, recirculation filter quantity, and wind penetration.

---

[1]Mining engineer, Pittsburgh Research Laboratory, National Institute for Occupational Safety and Health, Pittsburgh, PA.

INTRODUCTION

Overexposure to airborne respirable crystalline silica (or quartz) dust can cause silicosis, a serious or fatal respiratory lung disease. Mining has some of the highest incidences of worker-related silicosis, and mining machine operators constitute the occupation most commonly associated with the disease [NIOSH 2003]. The Mine Safety and Health Administration (MSHA) enacts and enforces mine worker safety and health standards to mitigate mine worker injuries and occupational diseases.

MSHA's permissible exposure limit is 2.0 mg/m^3 of airborne respirable dust for coal mine workers as defined by the U.K. Mining Research Establishment (MRE) criteria [30 CFR[2] 70–72, 74 (2007)]. If more than 5% quartz mass is determined to be in the coal mine worker dust sample using MSHA's P7 infrared method [Parobeck and Tomb 2000], the applicable respirable dust standard is reduced to the quotient of 10 divided by the percentage of quartz in the dust sample. MSHA's nuisance dust limit (total dust) for metal/nonmetal miners is 10 mg/m^3 as defined by the American Conference of Governmental Industrial Hygienists [ACGIH 1973; 30 CFR 56–58 (2007)]. If more than 1% quartz mass is determined to be in the metal/nonmetal mine worker dust sample using the National Institute for Occupational Safety and Health (NIOSH) X-ray Method [Parobeck and Tomb 2000], the applicable standard is then a respirable dust standard of 10 divided by the sum of the quartz percentage plus 2. Both of these dust standards are intended to limit worker respirable crystalline silica (quartz) exposure to 0.1 mg/m^3 or less for the shift.

Mine worker overexposure to quartz dust continues to be a problem at U.S. mining operations. The percentages of MSHA dust samples from 2000 to 2004 that exceeded the respirable dust standard due to quartz were 11% for sand and gravel mines, 11% for stone mines, 19% for nonmetal mines, 17% for metal operations, and 17% for coal mines [NIOSH 2008]. At surface mining operations, the occupations that have the highest frequency of exceeding the respirable dust standard are usually operators of mechanized excavation equipment, such as drills, bulldozers, scrapers, front-end loaders, haul trucks, and crushers [Tomb et al. 1995].

A primary means of dust control on mechanized surface mining equipment is enclosed operator cabs with an air filtration system. Field assessment of six surface coal mine rock drills and five bulldozers by NIOSH have shown that rock drill dust generation was one order of magnitude higher than bulldozer dust generation and that enclosed cab dust reduction efficiency for this equipment varied from 44% to nearly 100% [Organiscak and Page 1999]. This study further showed a wide variability in dust concentration and silica content within the same enclosed cab measured intermittently over an 8-month period [Organiscak and Page 1999]. Additional NIOSH field studies of retrofitting five older enclosed cabs with air filtration system improvements also showed their cab dust reduction efficiency varied from 64% to 99% [Chekan and Colinet 2003; Organiscak et al. 2003; Cecala et al. 2003, 2005]. These studies indicate that cab air filtration system design and operational factors influence dust control effectiveness and the ability to control operator dust exposure.

To better qualify air filtration system design and operational factor effects on enclosed cab dust control performance, controlled laboratory experiments were performed on an enclosed cab test stand at the NIOSH Pittsburgh Research Laboratory (PRL). These experiments examined the independent factor effects of intake filter efficiency, intake filter loading (airflow resistance), intake air leakage around the filter, recirculation filter use, and wind on cab performance.

---

[2]*Code of Federal Regulations.* See CFR in references.

The dependent cab performance variables measured included cab particulate penetration, intake airflow, recirculation airflow, intake filter pressure, cab pressure, and intake air leakage. Additional experiments were also conducted on the enclosed cab test stand to investigate the effects of adding an intake pressurizer to the filtration system.

TEST APPARATUS AND MEASUREMENT METHODS

An experimental cab test apparatus was constructed having cab filtration system features similar to those of existing equipment cabs. The cab test apparatus was a 72-ft^3 painted plywood enclosure 6 ft high by 3 ft wide by 4 ft deep on rolling casters (Figure 1). The front side was a hinged door with a Plexiglas window to observe the interior of the enclosure. The enclosure joints were sealed with silicon, and the entry door was sealed with high-density foam tape to ensure good cab integrity. Three 1-in-diam holes were uniformly spaced in the Plexiglas window on the front door and on the opposing back side wall of the cab to allow intake air to uniformly exit the cab at positive pressure.

Figure 1.—Experimental cab test apparatus.

A 27.6-V dc, variable-speed, Ametek RTP1400 brushless dual-fan blower was mounted on the front half of the enclosure roof with discharge vents located through the cab ceiling near the front door. The dual-fan blower's air pressure-quantity characteristic curve at maximum speed is shown in Figure 2. A mockup roof-mounted HVAC Plexiglas housing encased the dual-fan blower, and a 1-ft by 2-ft cab recirculation air inlet was placed through the opposing back side of the roof/ceiling. A frame and holding bracket were incorporated around the ceiling inlet for installing a pleated panel filter. Another 1-ft by 2-ft cab inlet was placed near the back floor of the cab and was connected to the back side of the mockup HVAC enclosure on the cab roof with a transition, two 90° PVC elbows, and 6-in-diam PVC pipe. An inlet cover panel with high-density foam on the perimeter was used to close either inlet during testing. During this testing, the cab recirculation air was drawn only through the ceiling inlet, which is similar to many of the roof-mounted retrofit HVAC systems.

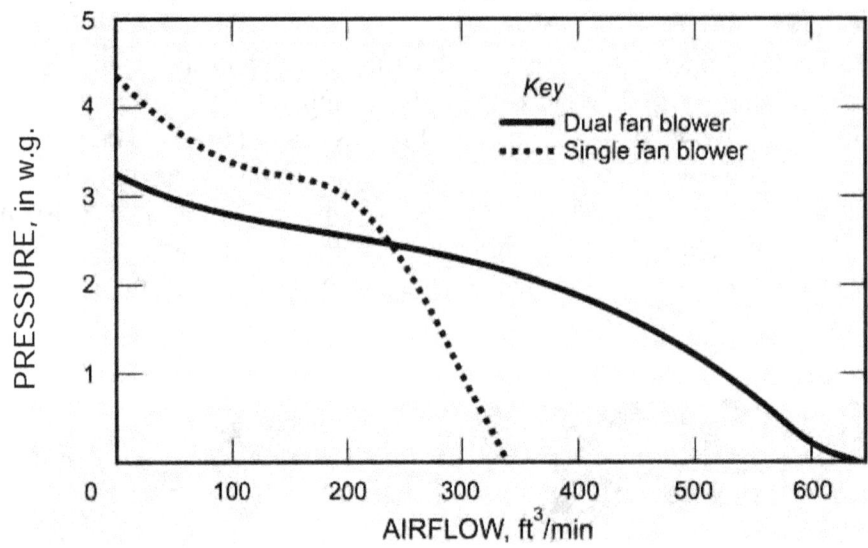

Figure 2.—Air pressure-quantity characteristic curves of fans on experimental cab apparatus.

Outside makeup air was brought into the side of the mockup HVAC system housing through either of two 3-in-diam PVC pipes connected to an exterior Plexiglas filter box. One of the pipes drew air from the filter box with only the recirculation fans. The other pipe could be pressurized with intake air from a 15- to 27.6-V dc, variable-speed, Ametek ECDC brushless single-fan blower located inside the filter box. The single-fan blower's air pressure-quantity characteristic curve at maximum speed is shown in Figure 2. Both PVC intake air pipes were fitted with ball valves so either intake delivery system could be individually tested. The filter sampling box had an inlet hole and bracket to accommodate an intake cylindrical filter cartridge on the exterior of the box. The filter box also had a ½-in-inside-diam barbed hose fitting opening for leakage testing around the intake filter.

Several of the cab's operating parameters were measured during testing with static air pressure gauges and airflow monitors, electronically recording to a Telog R-3307 seven-channel data acquisition system (Telog Instruments, Inc., Victor, NY). The negative differential pressure across the exterior to interior of the intake filter box was measured with a 0- to 2-in w.g. Magnehelic pressure instrument with a 4- to 20-mA output (Dwyer Instruments, Inc., Michigan City, IN). Cab enclosure positive-pressure differential was measured with a 0.0- to 0.5-in w.g. Magnehelic pressure instrument with a 4- to 20-mA output (Dwyer Instruments, Inc., Michigan City, IN). Leakage into the filter box was measured with a 0- to 300-L/min TSI Model 4040 Thermal Mass Flowmeter with a 0- to 10-V analog output (TSI, Inc., Shoreview, MN). Wind speed was measured on the top left corner of the cab with a 0- to 6,000-ft/min AIRFLOW AV6 Digital Handheld Vane Anemometer with a 0- to 1-V analog output to verify consistent airflow conditions during the test (AIRFLOW, Buckinghamshire, U.K.).

Other cab operating data measured before and after each test were intake airflow, recirculation airflow, average wind speed, and atmospheric conditions. Intake airflow velocity was centerline measured inside the 3-in-diam PVC intake pipe with a 0- to 6,000-ft/min TSI Model 8346 VelociCALC Hot Wire Anemometer (TSI, Inc., Shoreview, MN). The recirculation airflow was measured with a 0- to 2,000-ft^3/min Alnor Standard Balometer Capture Hood placed over the ceiling inlet/filter (TSI, Inc., Alnor Products, Shoreview, MN). Wind speed measurements were made with a Davis handheld vane anemometer for 1-min periods on each side and top of the cab (Figure 3). Atmospheric wet- and dry-bulb temperatures were taken with a Davis Inotek battery-operated psychrometer (Davis Inotek, Baltimore, MD). Barometric pressure was measured with a Pretel AltiPlus K2 Electronic Altimeter (France).

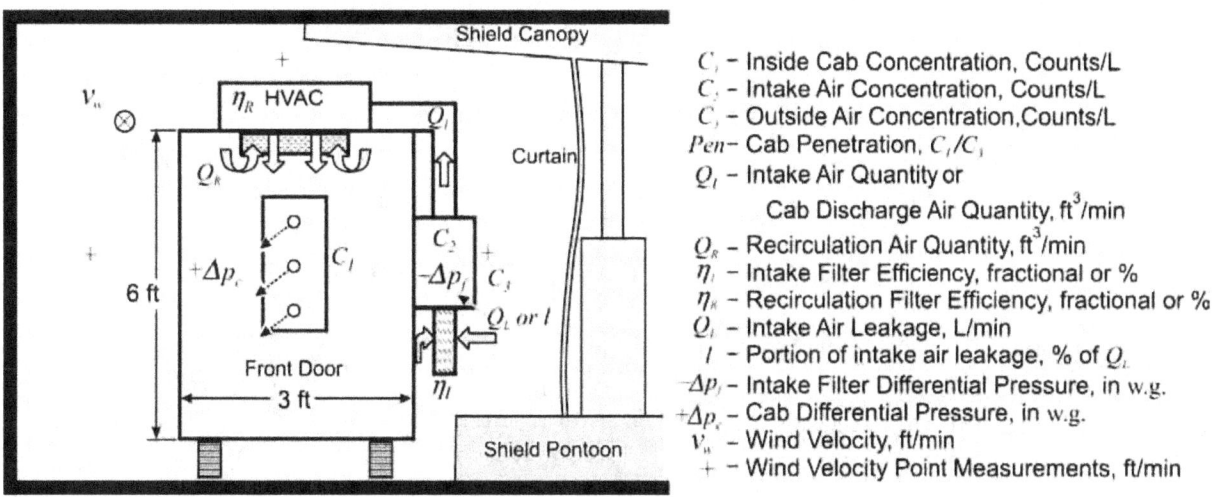

Figure 3.—Laboratory cab test apparatus used in PRL's longwall test gallery.

The cab particulate penetration performance was measured by relative comparisons of particle count concentrations inside (C_1) and outside (C_3) the cab test stand, challenged with ambient air particles (see Figure 3). Portable handheld HHPC-6 particle counters with six custom channel sizes of 0.3, 0.5, 0.7, 1.0, 3.0, and 5.0 μm were operated at 2.83 L/min (0.1 ft^3/min) (Hach Ultra Analytics, Grants Pass, OR). Differential size particle counting was conducted in concentration mode over a sample volume of 2.83 L or for 1-min sampling periods. The instruments were mounted inside the enclosure and sampled at the designated locations remotely through 18-in lengths of 1/8-in-inside-diam Tygon tubing with isokinetic inlet probes. The manufacturer's 0.45-in-diam isokinetic inlet probes were used at all locations except on the outside sampling location during the wind tests. For these tests, a 1/8-in-diam isokinetic probe inlet was used to more closely match up the inlet sampling velocity to the incoming wind velocity. Particle counts per liter were recorded for 1-min time periods in the instruments' internal buffer/memory. Since the largest measurable fraction of ambient air particles was in the submicron size range, the three smaller particle counter channels were summed to determine the submicron (0.3- to 1.0-μm) respirable particle count concentrations inside (C_1) and outside (C_3) the cab enclosure for each minute of the test. Also, cab intake air particle count concentrations

(C_2) were measured with another HHPC-6 inside the filter box to determine intake filter efficiency under no leakage conditions around the intake filter.

Submicron particle cab penetration ($Pen = C_1/C_3$) performance was determined from corresponding 15-min averages at reasonably stable interior cab concentrations. The time for the cab enclosure concentrations to decay and reach interior stability depended on several factors such as intake filter efficiency, intake airflow, recirculation filter use, initial inside particle count concentration, and outside particle count concentration. One presumption for interior cab concentration stability is a constant or stable outside concentration. Preliminary cab testing indicated that after closing the enclosure door most of the interior concentration decay occurred within 15 and 30 min with and without the recirculation filter, respectively. Ambient air concentrations were also found to be reasonably stable during these preliminary tests. Therefore, experimental cab tests were conducted for 30- and 45-min periods with and without the recirculation filter, respectively, to achieve a reasonably steady concentration averaging period for the last 15 min of a test. A cab decay time for each test was estimated by the number of 1-min time periods it took to reach the average inside concentration for the last 15 min of the test. Finally, it must be noted that cab penetration (Pen) will be reported throughout this report, but can be easily converted to a cab reduction efficiency (% cab reduction efficiency = $(1 - Pen) \times 100\%$) or a cab protection factor (cab protection factor = $1/Pen$) [Organiscak et al. 2003].

EXPERIMENTAL CAB TEST FACTORS

Experiments were conducted on the cab test apparatus to study multiple filtration system factors on cab penetration. Table 1 shows these experimental test factors for cab filtration systems without and with an intake pressurizer fan (referred to as "pressurizer"). The test factors studied on the cab filtration system without the pressurizer were intake filter efficiency, intake filter loading (airflow resistance), intake air leakage around the filter, recirculation filter use, and wind. This series of testing was conducted in PRL's longwall test gallery with the cab's front door and three air exit holes oriented into the wind direction, as shown in Figure 3. The cab was positioned in the cross-section of the gallery so as to achieve reasonably equal air velocities on both sides and top of the cab. The maximum wind velocity that could be reached inside the longwall gallery was 10 mph. Wind infiltration into the cab was previously shown to occur when cab pressure is exceeded by wind velocity pressure [Heitbrink et al. 2000].

Table 1.—Experimental cab test factors

Test factors	Filtration system without intake pressurizer fan		Filtration system with intake pressurizer fan	
	Low-level (−1)	High-level (+1)	Low-level (−1)	High-level (+1)
(A) Intake filter efficiency	Single-stage	Multistage (a)	Single-stage	Multistage (a)
(B) Intake filter loading	Unloaded	Loaded (b)	Unloaded	Loaded (b)
(C) Intake air leakage	Sealed	½-in hole (c)	Sealed	½-in hole (c)
(D) Recirculation filter	None	Panel filter (d)	None	Panel filter (d)
(E) Wind	Calm	10 mph (e)	Calm	Calm

Identical experimental test factors were studied on the cab filtration system with the pressurizer, except for wind. Wind was excluded from the pressurizer tests since the cab pressure was certain to be above the 0.05-in w.g. velocity pressure generated by the 10-mph wind velocity inside the longwall test gallery. This series of testing was conducted in the high bay area outside the gallery, as shown in Figure 1.

The experimental cab test factors shown in Table 1 are described below. The low- and high-level conditions are mathematically represented by −1 and +1, respectively, for subsequent linear regression modeling of the test levels. The high level of cab test factors A, B, C, D, and E in Table 1 are also coded by lower-case letters a, b, c, d, and e, respectively, to conveniently describe test conditions. For example, the test condition *ade* without the pressurizer represents a multistage intake filter (a), an unloaded intake filter, a sealed intake leakage, a recirculation panel filter (d), and a 10-mph wind velocity (e) test.

(*A*) Intake Filter Efficiency

- <u>Low-level (−1)</u>: A single-stage, round pleated cellulose filter cartridge (7-in-diam by 13-in-long, Donaldson Co., Inc., Minneapolis, MN) with lower submicron particle size filter efficiency.
- <u>High-level (+1) (*a*)</u>: A multistage, round microglass and electrostatic contiguous layered filter cartridge (7-in-diam by 12-in-long, Clean Air Filter, Defiance, IA) with higher submicron particle size filter efficiency.

(*B*) Intake Filter Loading

- <u>Low-level (−1)</u>: An unloaded intake filter was tested in what was considered as new condition (without any exposure to heavy or coarse dust loading).
- <u>High-level (+1) (*b*)</u>: A loaded intake filter was simulated by placing a round cut piece of 14-gauge perforated plate (3/32-in-diam holes staggered 3/16 in center to center) fitted flush within the interior of the filter gasket area and outlet hole of the filter cartridge. A 2-in-wide strip of duct tape was also placed down the center of the perforated plate to help noticeably increase filter resistance. Increasing intake filter resistance is used to simulate dust-loading effects on the cab filtration system.

(*C*) Intake Air Leakage

- <u>Low-level (−1)</u>: The ½-in-inside-diam hole in the filter box was sealed or closed.
- <u>High-level (+1) (*c*)</u>: The ½-in-inside-diam hole in the filter box was open. The TSI Model 4040 Thermal Mass Flowmeter was connected with tubing to this hole for measuring the quantity of the leak.

(*D*) Recirculation Filter

- <u>Low-level (−1)</u>: None used. A 12-in-wide by 24-in-long by 4-in-deep 2×4 wood-constructed open-filter frame blank was inserted into the aluminum frame filter holding bracket with a rectangular perforated restrictor plate (same material used for loading the intake filter) covering the inlet area side of the bracket. The restrictor plate had equally spaced 2-in-wide duct tape strips across it to achieve a targeted balance of 25 ft^3/min of intake air for the unloaded and more restrictive Clean Air Filter intake filter when used

without the recirculation filter and pressurizer. The HVAC dual-fan blower had to be run at maximum speed to achieve this target intake airflow.
   • High-level (+1) (d): The recirculation filter used was an American Air Filter (AAF) pleated microglass panel filter (12-in-width by 24 in-length by 4-in-depth nominal size). It had an American Society of Heating, Refrigerating and Air-Conditioning Engineers (ASHRAE) minimum efficiency reporting value (MERV) of 15, or 85%–94.9% in the 0.3- to 1.0-μm size range at a rated airflow capacity of 1,000 ft^3/min. This filter was inserted into the aluminum frame holding bracket with the perforated restrictor plate.

(E) Wind (only tested on the cab filtration system without pressurizer)
   • Low-level (−1): Cab was tested at a calm air velocity condition inside the longwall test gallery.
   • High-level (+1) (e): Cab was tested at a 10-mph wind velocity condition inside the longwall gallery.

Cab filtration system fan speeds were kept constant throughout experiments to examine the test factor effects on cab performance. All of the tests were conducted with the HVAC dual-fan blower set to maximum speed. The intake pressurizer testing was conducted with its fan speed set in the middle of its operating range so the cab pressure instrumentation would not exceed its maximum of 0.5 in w.g.

EXPERIMENTAL DESIGN

Experiments were conducted on all of the cab test factor combinations shown in Table 1 for each filtration system. These cab factor test combinations were conducted in several series or blocks of experiments. The first series of experiments was conducted on the cab filtration system without the intake pressurizer. Laboratory testing of this filtration system configuration was based on a five-factor, two-level factorial experimental design [Myers and Montgomery 1995]. This design was split into two blocks of half-fraction experiments (see Appendix A) [Myers and Montgomery 1995]. Each half-fraction is a full two-level factorial design for the four cab factor configurations (*ABCD*) with wind velocity (*E*) testing split equally between the half-fraction blocks of experiments. This design permits screening of a half-fraction block of data for the significant single factor and two factor interactions [Myers and Montgomery 1995].

The experimental run conditions were randomized, but testing was conducted by running a test period with one HHPC-6 instrument sampling inside and another HHPC-6 instrument sampling outside the cab enclosure and then switching these instruments for a subsequent second test period under the same experimental run conditions. Each experimental run condition was randomly conducted twice, providing four enclosed cab testing periods. Although the particle counting instruments were individually factory-calibrated, they were switched for the subsequent test periods to average out any instrument biases. Experimental runs were usually repeated more than two times if the ambient test concentration exceeded 100,000 counts/L or if there was noticeable cab penetration variation (standard deviation > 0.035). Since preliminary statistical analysis on the first half-fraction block of the experimental design indicated significance for all factors either individually or as interactions, the second half-fraction block of the experimental design was subsequently conducted to complete the full five-factor, two-level factorial experimental design. A total of 74 randomized conditional runs or 148 tests were conducted for the complete two-level factorial experimentation. The first and second half-fraction of experimental data are shown in Tables B-1 and B-2, respectively, in Appendix B.

Lastly, another series or block of experiments was conducted on the cab test apparatus configured with the pressurizer fan. These tests were conducted without wind and in similar fashion as described above. This testing followed the four-factor, two-level factorial experimental design (*ABCD*) shown in Table A-1. A total of 34 randomized conditional runs or 68 tests were completed during these experiments. Table B-3 shows the pressurizer block of experimental data.

EXPERIMENTAL RESULTS

The two largest test factors that influenced cab penetration (*Pen*) for all of the experiments were intake filter efficiency and recirculation filter. Figures 4 and 5 show box-and-whisker plots of the cab penetration data classified by the intake filter and recirculation filter use for the first series of experiments without the pressurizer and for the second series of experiments with the pressurizer, respectively. Each box-and-whisker section represents 25% of the data collected, with the median displayed in the middle of the boxes. The open point shown outside the whisker in Figure 4 is an outlying data point. Figures 4 and 5 illustrate significant differences in cab *Pen* between the intake filters by themselves and with a recirculation filter. Using the recirculation filter made a significant reduction in cab *Pen* compared to the intake filter by itself. The figures also show that the cab *Pen* performance of the lower-efficiency intake filter in combination with the recirculation filter was similar to the cab performance of the higher-efficiency intake filter by itself. The effects of the other experimental test factors can be seen in the spread of *Pen* data in both of these figures.

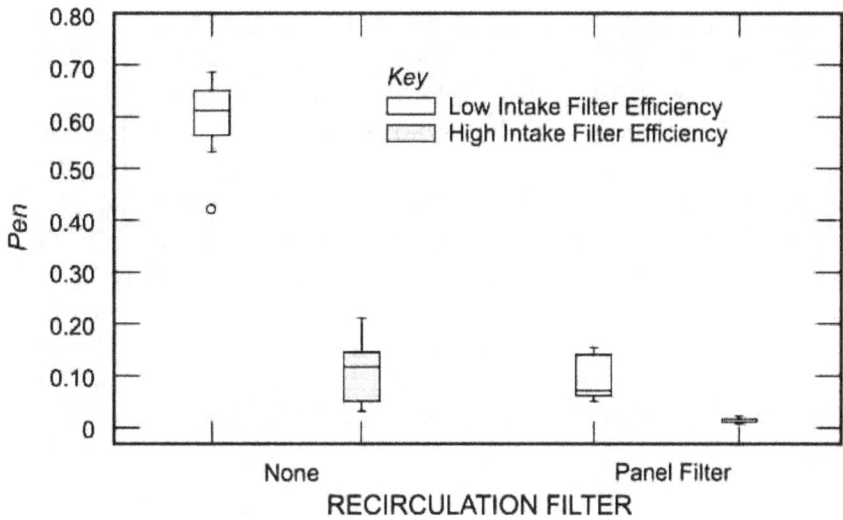

Figure 4.—Box-and-whisker plot of cab *Pen* for filter combinations without pressurizer.

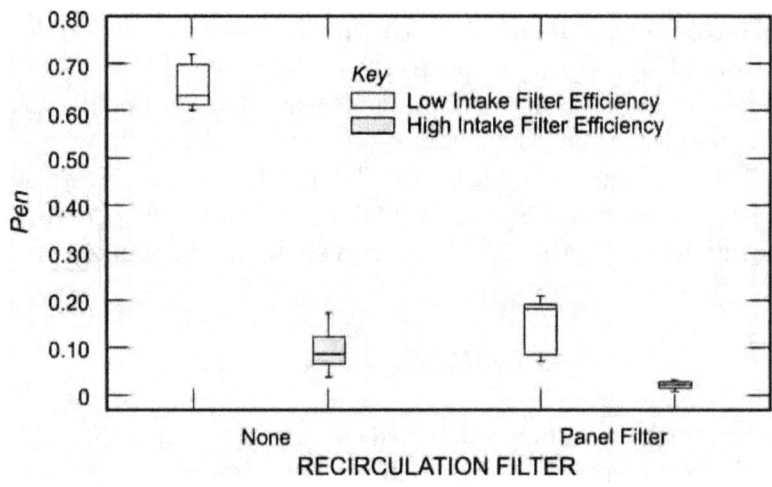

Figure 5.—Box-and-whisker plot of cab *Pen* for filter combinations with pressurizer.

Cab *Pen* and other cab performance statistics were also computed and examined with respect to the experimental test factors. Table 2 shows the cab performance statistics (average and minimum-maximum) for three key test factors (intake filter efficiency (*A*), intake filter loading (*B*), and recirculation filter (*D*)) for the first series of experiments on the filtration system without a pressurizer. Average intake filter efficiencies measured for submicron size particles (0.3–1.0 μm) are also reported in this table. The wind velocity conditions were not differentiated in this table since this factor did not exhibit noticeable differences in cab *Pen* compared to the other experimental factors. Table 3 shows similar cab performance statistics for the second series of cab experiments on the filtration system with a pressurizer.

Table 2 again shows that the largest reductions in cab *Pen* without the pressurizer were achieved with an increase in intake filter efficiency and the use of a recirculation filter. The lower-efficiency filter provided an average cab *Pen* of 0.635 and 0.569 for the unloaded and loaded intake filter, respectively, without the recirculation filter. These average cab *Pens* significantly decreased to 0.134 and 0.054, respectively, with the recirculation filter. The higher-efficiency filter provided an average cab *Pen* of 0.072 and 0.131 for the unloaded and loaded intake filter, respectively, without the recirculation filter. These average *Pens* significantly decreased to 0.007 and 0.009, respectively, with the recirculation filter. The recirculation filter also decreased the decay time needed for the cab interior concentrations to go down and stabilize after the cab door was closed. The average decay times ranged from 16 to 29 min without the recirculation filter and from 7 to 9 min with the recirculation filter.

Table 3 similarly shows that the largest reductions in cab *Pen* with the pressurizer were achieved with an increase in intake filter efficiency and the use of a recirculation filter. The lower-efficiency filter provided an average cab *Pen* of 0.693 and 0.609 for the unloaded and loaded intake filter, respectively, without the recirculation filter. These average *Pens* significantly decreased to 0.194 and 0.073, respectively, with the recirculation filter. The higher-efficiency filter provided an average cab *Pen* of 0.071 and 0.108 for the unloaded and loaded intake filter, respectively, without the recirculation filter. These average *Pens* significantly decreased to 0.009 and 0.010, respectively, with the recirculation filter. The recirculation filter also decreased the decay time needed for the cab interior concentrations to go down and stabilize after the cab door was closed. The average decay times ranged from 17 to 25 min without the recirculation filter and from 6 to 11 min with the recirculation filter.

Table 2.—Cab testing results without pressurizer
(Top number is average; bottom *italicized* numbers are minimum-maximum range)

Intake filter and efficiency, % (A)	Intake filter loading (B)	Recirculation filter (D)	Pen C_1/C_3	Q_I, ft³/min	$-\Delta p_f$, in w.g.	I % of Q_I	Q_R, ft³/min	$+\Delta p_c$, in w.g.	Decay time, min
Single-stage 35%	Unloaded	None	0.635 *0.557-0.690*	48.8 *45.4-50.6*	0.16 *0.14-0.18*	0.8 *0.0-1.7*	358 *338-368*	0.24 *0.21-0.28*	16 *1-38*
Single-stage 32%	Unloaded	Panel filter	0.134 *0.122-0.148*	58.7 *56.0-61.0*	0.22 *0.19-0.23*	0.8 *0.0-1.8*	318 *300-328*	0.31 *0.28-0.37*	7 *1-21*
Single-stage 44%	Loaded	None	0.569 *0.426-0.637*	21.5 *20.5-22.3*	0.50 *0.46-0.53*	3.7 *0.0-7.8*	378 *368-390*	0.08 *0.05-0.12*	18 *3-38*
Single-stage 42%	Loaded	Panel filter	0.054 *0.045-0.059*	25.2 *23.6-27.2*	0.69 *0.67-0.72*	4.3 *0.0-7.9*	337 *332-345*	0.09 *0.06-0.10*	9 *1-23*
Multistage >99%	Unloaded	None	0.072 *0.027-0.132*	22.8 *21.0-25.0*	0.48 *0.45-0.51*	3.4 *0.0-7.1*	383 *370-390*	0.09 *0.06-0.12*	27 *15-36*
Multistage >99%	Unloaded	Panel filter	0.007 *0.002-0.012*	28.7 *26.2-30.2*	0.64 *0.62-0.67*	3.2 *0.0-6.5*	332 *318-345*	0.10 *0.07-0.12*	7 *2-20*
Multistage >99%	Loaded	None	0.131 *0.040-0.211*	14.9 *13.8-16.2*	0.54 *0.50-0.58*	3.7 *0.1-11.6*	388 *365-398*	0.06 *0.03-0.09*	29 *12-39*
Multistage >99%	Loaded	Panel filter	0.009 *0.003-0.014*	18.8 *17.2-20.2*	0.74 *0.71-0.77*	6.3 *0.1-10.8*	344 *330-350*	0.06 *0.04-0.09*	9 *1-23*

Table 3.—Cab testing results with pressurizer
(Top number is average; bottom *italicized* numbers are minimum-maximum range)

Intake filter and efficiency, % (A)	Intake filter loading (B)	Recirculation filter (D)	Pen C_1/C_3	Q_I, ft³/min	$-\Delta p_f$, in w.g.	I % of Q_I	Q_R, ft³/min	$+\Delta p_c$, in w.g.	Decay time, min
Single-stage 29%	Unloaded	None	0.693 *0.636-0.720*	80.1 *78.2-82.0*	0.31 *0.31-0.33*	0.8 *0.0-1.6*	342 *340-348*	0.44 *0.42-0.45*	22 *0-36*
Single-stage 29%	Unloaded	Panel filter	0.194 *0.179-0.211*	91.8 *89.4-93.4*	0.39 *0.38-0.40*	0.9 *0.0-1.6*	310 *305-315*	0.47 *0.44-0.49*	8 *1-26*
Single-stage 39%	Loaded	None	0.609 *0.596-0.620*	30.2 *29.2-31.4*	0.96 *0.94-1.01*	3.8 *0.0-7.7*	383 *370-395*	0.10 *0.09-0.11*	17 *3-40*
Single-stage 39%	Loaded	Panel filter	0.073 *0.064-0.079*	33.2 *31.9-34.8*	1.16 *1.13-1.21*	3.8 *0.0-7.7*	338 *332-345*	0.12 *27-32*	11 *1-21*
Multistage >99%	Unloaded	None	0.071 *0.030-0.107*	39.2 *38.0-40.8*	0.87 *0.84-0.88*	2.8 *0.0-5.7*	370 *358-378*	0.16 *0.14-0.17*	25 *12-36*
Multistage >99%	Unloaded	Panel filter	0.009 *0.004-0.012*	44.8 *43.4-46.0*	249 *0.99-1.02*	2.7 *0.0-5.4*	335 *325-342*	0.20 *0.18-0.21*	8 *2-21*
Multistage >99%	Loaded	None	0.108 *0.037-0.178*	23.1 *21.4-25.0*	1.04 *1.02-1.06*	4.0 *0.1-10.0*	387 *380-395*	0.07 *0.06-0.08*	20 *13-32*
Multistage >99%	Loaded	Panel filter	0.010 *0.003–0.018*	26.4 *24.6-28.6*	1.24 *1.23-1.25*	4.9 *0.1-9.8*	341 *330-350*	0.08 *0.07-0.09*	6 *1-16*

Adding the intake pressurizer fan to the cab filtration system resulted in minor changes to the cab *Pen* from the increased airflow through the intake filter. A comparison of Tables 2 and 3 shows the cab *Pen* for the lower-efficiency intake filter tests perceptibly increased with the addition of the pressurizer. This corresponded to higher intake airflows and decreased intake filter efficiency with the pressurizer versus without the pressurizer. Cab *Pen* change was negligible for the higher-efficiency filter with the addition of the pressurizer, corresponding to negligible changes in intake filter efficiency over the range of airflows achieved with and without the pressurizer. The pressurizer did not significantly change the recirculation airflow quantity (Q_R) for identical filter combinations.

The intake filter differential pressure ($-\Delta p_f$), cab intake airflow quantity (Q_I), cab differential pressure ($+\Delta p_c$), and intake air leakage (l) all noticeably changed for the filter test factor combinations and pressurizer as shown in Tables 2 and 3. Figure 6 illustrates the indirect relationships between intake filter differential pressure ($-\Delta p_f$) and cab intake airflow quantity (Q_I) for these experiments. The intake filter differential pressure and airflow quantity data are grouped by recirculation filter and pressurizer use, with group associations indicated by dashed lines. The data show that the differential pressure across the intake filter was inversely related to intake air quantity for all data groups. Adding a recirculation filter increased both the intake airflow and intake filter differential pressure, shifting the associated relationship to the top right of the graph. The pressurizer additionally increased the intake airflow and filter differential pressure, further shifting these associated relationships to the top right of the graph.

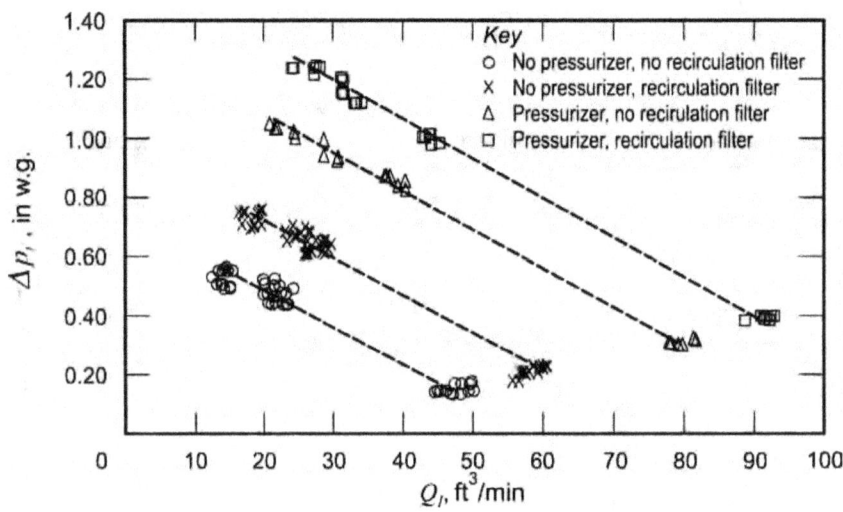

Figure 6.—Relationship between intake filter differential pressure and intake airflow.

Figure 7 shows the direct cab differential pressure ($+\Delta p_c$) relationship with respect to intake air quantity (Q_I), with points classified by wind and pressurizer use. This figure clearly indicates the direct relationship between cab pressure and intake air quantity. It also shows that wind increased cab differential pressure by roughly the wind velocity pressure. The pressurizer further increased the intake air quantity and cab pressure.

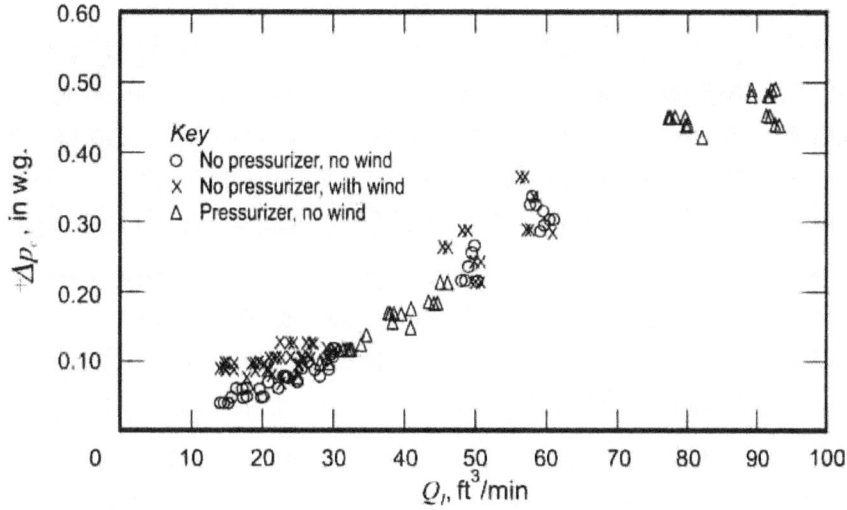

Figure 7.—Relationship between cab differential pressure and intake airflow.

The relationship between leakage (l) and intake filter differential pressure ($-\Delta p_f$) with the ½-in-diam leakage hole open is shown in Figure 8. The leakage data are categorized by recirculation filter and pressurizer use with dashed lines drawn through these data groups to illustrate their associations. This figure shows the direct relationship between intake leakage and filter differential pressure for all of the data groups. The higher-efficiency intake filter and loading conditions increased the differential pressure and leakage across all data groups.

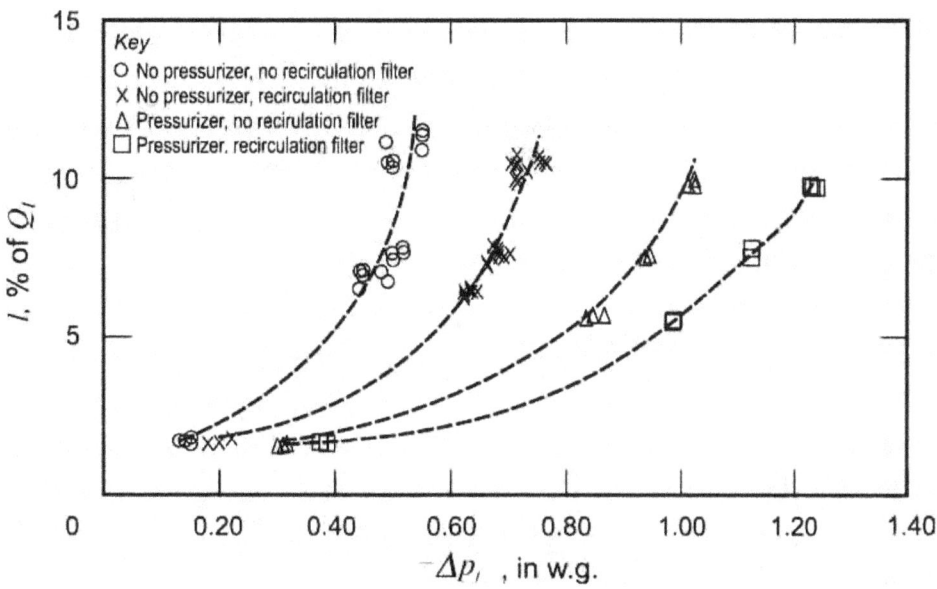

Figure 8.—Relationship between intake leakage and intake filter differential pressure.

STATISTICAL ANALYSIS

Linear regression analysis was conducted to statistically quantify the relationship between the cab testing factors (considered independent variables) and cab penetration (dependent variable). Since the box-and-whisker plots of Figures 4 and 5 illustrate an extensive data range, nonnormality, and unequal variance in the dependent cab penetration variable, it was transformed by using natural logarithms (ln *Pen*) to stabilize regression modeling variance [Myers and Montgomery 1995]. Linear regression analysis was conducted on comparative sets of experimental data. Both half-fractions of the cab filtration configuration without the pressurizer were analyzed together for the five experimental test factors (intake filter efficiency (*A*), intake filter loading (*B*), intake air leakage (*C*), recirculation filter (*D*), and wind (*E*)). A stepwise regression analysis of the dependent variable (ln *Pen*) with respect to the single factors and two-factor interactions was conducted and is shown in Appendix C. The cab filtration system with and without the intake pressurizer configurations were also comparatively analyzed excluding the 10-mph wind tests. Appendix D shows this stepwise regression model and analysis for the dependent variable (ln *Pen*) with respect to the single factors, two-factor interactions, and pressurizer (a blocking factor). All stepwise regression model factors and interactions were successively selected by the highest level of significance on cab *Pen* with no factor removal during the analysis. Table 4 shows the statistically significant experimental factors and interactions, listed in descending order of significance for both regression analyses.

Table 4.—List of statistically significant regression factors affecting cab *Pen*

Regression selection order	Filtration system tests without intake pressurizer fan	Filtration system tests with and without intake pressurizer fan, excluding wind
1	Recirculation filter (*D*)	Intake filter efficiency (*A*)
2	Intake filter efficiency (*A*)	Recirculation filter (*D*)
3	Intake filter efficiency × loading (*AB*)	Leakage (*C*)
4	Leakage (*C*)	Intake filter efficiency × loading (*AB*)
5	Intake filter efficiency × leakage (*AC*)	Intake filter efficiency × leakage (*AC*)
6	Intake filter efficiency × recirculation filter (*AD*)	Loading × recirculation filter (*BD*)
7	Loading × recirculation filter (*BD*)	Intake filter efficiency × recirculation filter (*AD*)
8	Loading × wind (*BE*)	Loading (*B*)
9	Loading × leakage (*BC*)	Pressurizer (*P*)

Analysis of both filtration systems showed comparable top seven regression factors, selected in a somewhat different order. Table 4 again illustrates that the top two experimental factors were the intake filter efficiency and the recirculation filter for both filtration systems tested. It also shows that leakage had a significant effect on cab *Pen* for both systems. Furthermore, cab *Pen* was significantly affected by intake filter efficiency interactions with loading, leakage, and recirculation filter use. Wind (eighth, as an interaction) and the pressurizer (ninth, as a blocking factor) were some of the least significant factors selected by regression analyses. The analyses clearly reveal the multifaceted experimental factor effects on cab *Pen*, with interactions suggesting factor codependence with other cab operating variables.

MATHEMATICAL MODEL FOR CAB PENETRATION

The cab test factor codependence with several other cab operating variables can be observed in Tables 2 and 3 and Figures 6 through 8. The air pressure and quantity relationships for the cab filtration system indicate that air quantity balance of contaminants within the system may better describe cab penetration. This cab filtration mathematical model is developed in Appendix E. It was formulated from a basic time-dependent mass balance model of airborne substances within a control volume. This mathematical model was particularly formulated for steady-state conditions in Appendix E and is shown below. It describes cab penetration in terms of intake filter efficiency, intake air quantity, intake air leakage, recirculation filter efficiency, recirculation filter quantity, and outside wind quantity infiltration into the cab.

$$Pen = \frac{x}{C} = \frac{Q_I(1-\eta_I + l\eta_I) + Q_w}{Q_I + Q_R\eta_R} \qquad (E-12)$$

This equation can also be expressed in other useful forms:

$$Pen = \frac{1 - \eta_I + l\eta_I + \dfrac{Q_w}{Q_I}}{1 + \dfrac{Q_R}{Q_I}\eta_R} \qquad (E-13)$$

$$\text{or} \quad Pen = \frac{1 - \eta_I + \dfrac{Q_L}{Q_I}\eta_I + \dfrac{Q_w}{Q_I}}{1 + \dfrac{Q_R}{Q_I}\eta_R} \qquad (E-14)$$

where
- x = inside cab contaminant concentration,
- C = outside cab contaminant concentration,
- Pen = ratio of inside to outside contaminant concentration, or x/C,
- Q_I = intake air quantity into the cab,
- η_I = intake filter efficiency, fractional,
- Q_L = air leakage quantity around the intake filter,
- l = fractional portion of intake air leakage, or Q_L/Q_I,
- Q_R = recirculation filter airflow,
- η_R = recirculation filter efficiency, fractional,

and
- Q_w = wind quantity infiltration into the cab.

NOTE: The above equations are dimensionless, so air quantities used in these equations must have equivalent units. Also, filter efficiencies and intake air leakage used must be fractional values (not percentage values).

Verification of this model was examined using the experimental data. Many of the above model variables were directly measured during experimental testing, except for the recirculation filter efficiency and wind quantity infiltration into the cab. To determine the experimental recirculation filter efficiency for validating this model, additional particle counting testing was conducted upstream and downstream of the recirculation filter to measure its filter efficiency. The AAF recirculation filter panel was tested at the cab floor inlet with the ceiling inlet blocked, intake air ducts closed, and cab door opened. A sealed horizontal plywood barrier was installed inside the cab, 1.5 ft parallel and above the floor, to create a separate intake sampling duct section to the recirculation filter. A particle counter sampled the ambient air in the intake section to the filter, and a particle counter sampled the filtered air inside the middle straight section of the 6-in-diam PVC recirculation duct (see PVC tube on the outside of the cab shown in Figure 1). Isokinetic sampling inlets were used to match sampler inlet velocities to air duct velocities. A VelociCalc Hot Wire Anemometer was used to measure the airflow in the PVC recirculation duct, and the HVAC dual-fan blower was operated at full speed during these tests to achieve recirculation filter airflow comparable with the experiments. Filter tests were conducted over a 15-min sampling period with the particle counting instruments switched between successive tests.

Table 5 shows the filter efficiency results for eight filter tests. Average recirculation filter efficiency measured with ambient air was 72.4% for 346 ft^3/min of airflow. Another similar AAF filter panel (not used in the experiments) was tested and showed a comparable filter efficiency of 71.1% for 347 ft^3/min of airflow. These filter efficiencies were observed to be less than their MERV 15 rating of 85%–94.9% in the 0.3- to 1.0-μm size range. The lower filter efficiencies found in these particular tests are most likely due to a relatively larger portion of 0.3- to 0.5-μm-sized particles measured in ambient air compared to a more balanced aerosol size range used in the MERV test procedure.

Table 5.—Recirculation filter efficiency results for 0.3- to 1.0-μm-sized particles

Filter test	Test time, min	Filter airflow, ft^3/min	Upstream concentration, counts/L	Downstream concentration, counts/L	Filter efficiency,[1] %
1	15	346	11,296	2,859	74.7
2	15	346	10,912	2,738	74.9
3	15	348	24,934	7,007	71.9
4	15	347	24,020	6,723	72.0
5	15	344	28,718	8,216	71.4
6	15	346	39,395	11,382	71.1
7	15	344	36,635	10,427	71.5
8	15	346	30,505	8,702	71.5
Average	15	346	25,802	7,257	72.4

[1]Filter efficiency = ((upstream conc. − downstream conc.) / upstream conc.) × 100%.

Wind quantity infiltration into the cab during these experiments could not be directly measured, but could be estimated for the three orifice openings facing directly into the wind by applying the general orifice flow equation described in Appendix E (Equation E-15) and shown below [Streeter and Wylie 1979]. The particular orifice flow equation when the wind velocity pressure exceeds cab static pressure (Equation E-16) described in Appendix E could not be applied in these particular experiments because the wind velocity pressure never exceeded the cab pressure during the 10-mph wind tests [Heitbrink et al. 2000].

$$v_o = \frac{Q_o}{A_o C_d} = \sqrt{2\frac{\Delta p_o}{\rho_{air}}} \tag{E-15}$$

where v_o = air velocity through an orifice,
Q_o = airflow quantity through an orifice,
C_d = orifice discharge coefficient,
A_o = area of orifice,
Δp_o = air pressure differential across orifice,
and ρ_{air} = air density.

Wind infiltration into the cab was presumed to occur when wind velocity exceeded the cab exit air velocity out of the three orifices opposed to the wind. Cab exit air velocity was initially calculated by assuming that the measured intake air quantity into the cab would exit equally through the six 1-in-diam holes with a discharge coefficient of 0.61 (a reasonable circular orifice coefficient [Streeter and Wylie 1979], also used in the wind penetration Equation E-16 [Heitbrink et al. 2000]). When wind velocity exceeded this cab exit air velocity, the difference was anticipated to be the wind velocity penetration through the opposing front three holes in the cab door, with all of the cab airflow exiting out the back three holes. Equation E-15 was used to estimate this wind air quantity forced into the front three holes. Wind quantity infiltration into the cab was estimated to vary from 0.8 to 1.8 ft^3/min only for 10 tests under wind test conditions *abe* and *abce*. The high-efficiency intake filter (*a*) under loaded conditions (*b*) had some of the lowest cab intake airflows that could be overcome by the 10-mph wind (*e*) in these particular experiments, confirming the loading (*B*) and wind (*E*) regression factor interaction with cab *Pen* in Table 4. The increase in cab pressures measured for tests with wind versus without wind (see Figure 7) supports the premise that more airflow is disproportionately discharged out a smaller area through the back three holes of the cab.

The cab operating variables for these experiments were applied in the above mathematical penetration model to examine its agreement with the actual cab penetration measured by the particle counters. Intake filter efficiencies measured without leakage for the particular filter and loading conditions (shown in Tables 2 and 3) were used in the model, as well as the 72.4% recirculation filter efficiency determined above. The other model variables (intake air quantity, intake air leakage, recirculation air quantity, and wind quantity penetration) were obtained from the experimental data in Appendix B.

Figure 9 shows the graph for the mathematically modeled cab *Pen* results compared to the experimentally measured cab *Pen* for all test conditions. A unity line is drawn on the graph to visually inspect how well the model compared to the measurements. This figure illustrates that the model provides a reasonable estimate of the cab *Pen* using the cab filtration system operating variables. The open points in the lower left of the graph illustrate some of the *ab* tests conducted at lower outside particle count concentrations (<15,400 particle counts/L), noticeably increasing the measured cab *Pen*. Additional *ab* experimental tests were conducted to measure cab *Pen* at higher outside particle count concentrations (see Tables B-2 and B-3 in Appendix B). Others have also reported unreliable cab *Pen* measurements if outside cab particle counts are too low to show measurable differences with respect to those inside the cab [Heitbrink et al. 1998]. Given these experimental variations, the mathematical model seems to provide a reasonable estimate of the cab penetrations.

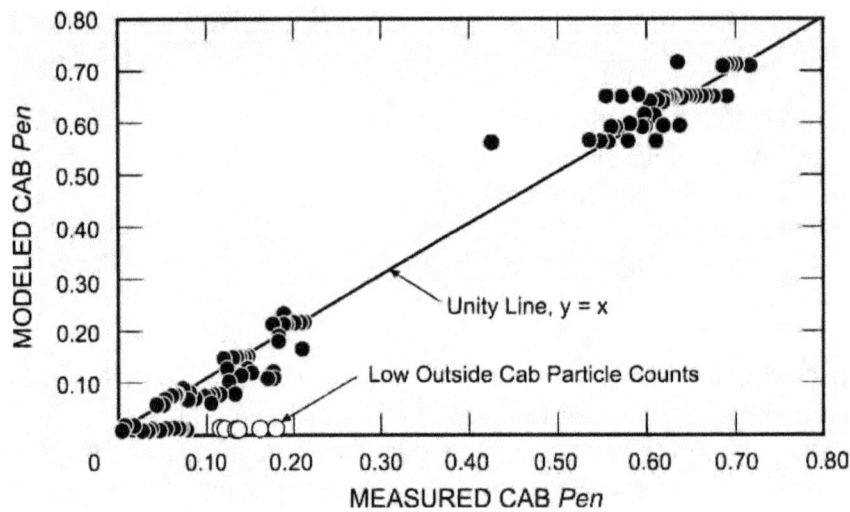

Figure 9.—Mathematically modeled cab *Pen* versus experimentally measured cab *Pen*.

DISCUSSION

The cab *Pen* measured in these laboratory experiments was conducted by counting particles found in ambient air. Figure 10 shows the size composition of ambient air particle concentrations measured for determining cab *Pen* during the last 15 min of all the tests. The three lines on the graph illustrate the 10th, 50th, and 90th percentiles for all of the experimental test data. Most of the particles were found in the submicron range, with a median (50th-percentile) submicron (0.3- to 1.0-μm) particle count concentration of 32,394 counts/L. These submicron particle count concentrations in the ambient air were found to remain reasonably constant during the cab test periods, but changed noticeably from day to day. Tables B-1, B-2, and B-3 in Appendix B illustrate this by the smaller differences observed in the average outside particle count concentrations (C_3) for the last 15 min of the test compared to the complete test and by the noticeably larger outside particle count concentrations between tests. This particle count variation is part of the experimental error.

The submicron particles in the ambient air were found to be a convenient and reasonable cab *Pen* test medium in these experiments. Only several particular run conditions were repeated because of higher cab *Pen* variability (standard deviation > 0.035) observed from lower ambient air particle count concentrations. Furthermore, coincidence error from high ambient particle concentrations (uncounted particles hidden behind other particles) seemed to be negligible in these experiments. The particle counter instrument coincidence error is specified at 5% for 2,000,000 particle counts/ft^3, or 70,670 particle counts/L. Only 25 of the 216 tests exceeded this concentration, with 7 tests exceeding 100,000 particle counts/L. Since most of the experimental runs having some of the higher ambient air concentration tests resulted in cab *Pen* standard deviations below 0.035, coincidence error was considered inconsequential.

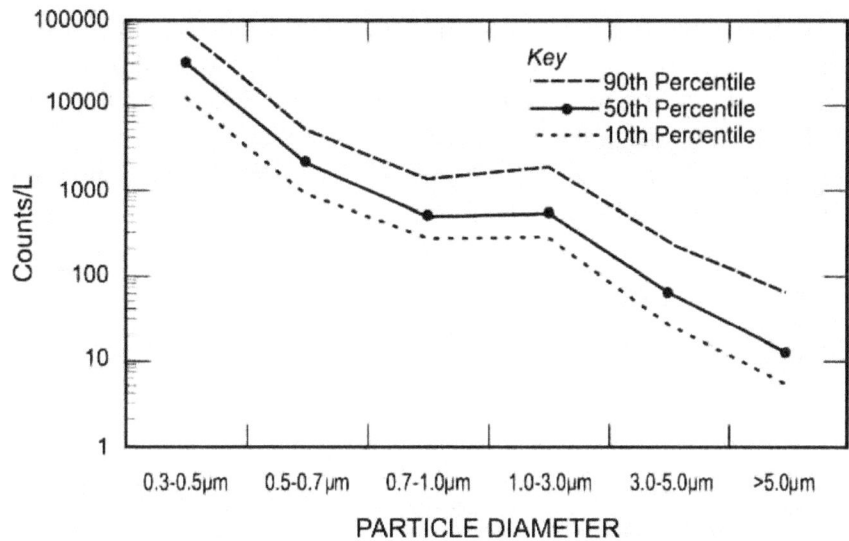

Figure 10.—Ambient air size classified particle count concentrations.

The experimental results and mathematical model developed have several limitations. First, the intake air leakage was placed in the negative air pressure plenum of the filtration system. Leakage into a negative air pressure plenum can be estimated by using the general orifice flow equation. If the leak is on the positive air pressure plenum of the filtration system, air leakage into the filtration system is very unlikely unless air velocity pressure is extremely high near the leak to induce air suction (venture effect) into the system. Secondly, experimental wind infiltration was more readily estimated from the general orifice equation because a portion of exit air was discharged directly into the wind. Wind infiltration through other cab exit air discharge configurations is much more difficult to determine and model. Wind infiltration into the cab for various discharge configurations can be minimized by maintaining cab pressure higher than wind velocity pressure. Finally, the mathematical model does not account for any internal cab contamination sources, such as transporting and dispersing contaminants inside the cab by the operator.

In addition, examining various levels of cab enclosure integrity was not part of this experimentation. These experiments were conducted on a laboratory cab test stand with reasonably tight, consistent, and well-controlled enclosure integrity. Exhaust air was discharged through three 1-in-diam holes on the front and rear of the cab test stand for a combined area of 0.033 ft^2, or 4.7 in^2. It was able to be pressurized to 0.10 in w.g. with 30 ft^3/min of intake air and had a minimum pressurization of 0.03 in w.g. with 13.8 ft^3/min of intake air under calm wind conditions (see Figure 7). Therefore, the cab pressure and exit air velocity were only low enough to be affected by the 10-mph wind velocity for a small subset of tests, when the higher-efficiency intake filter was under loaded conditions.

In previous NIOSH field studies, cabs were found to have varying degrees of enclosure integrity, indicated by their differences in cab pressures. In these field studies, five older enclosed cabs were retrofitted with air filtration system improvements, and their cab dust reduction efficiencies varied from 64% to 99% or cab penetrations of 0.36 to 0.01, respectively [Chekan and Colinet 2003; Organiscak et al. 2003; Cecala et al. 2003, 2005]. Table 6 summarizes these results in ascending order of performance achieved with these retrofitted installations. All of these cabs had a new roof-mounted HVAC unit installed with a pressurizer and filtration system. The Davey M8B drill, CAT 980B loader, and DrillTech DK40 drill had Red Dot roof-mounted HVAC systems and a Clean Air Filter intake filter pressurizer. The Euclid R-50 and IR DM45E drill had International Transit/Sigma HVAC systems with a single-stage fan pressurizer and a dual-fan pressurizer, respectively. Intake and recirculation filter efficiency performance specifications on the retrofitted cab systems were at least 95% on respirable-size dust. Pressurizer airflow specifications for these systems were equal or greater than 70 ft^3/min.

Table 6.—Summary of NIOSH enclosed cab field studies

Cab evaluation	Cab pressure, in w.g.	Wind velocity equivalent,[1] mph	Average inside cab dust level, mg/m^3	Average outside cab dust level, mg/m^3	Penetration, in/out
Davey M8B drill	None detected	0	0.08	0.22	0.36
Euclid R-50 truck	0.01	4.5	0.32	1.01	0.32
CAT 980B loader	0.015	5.6	0.03	0.30	0.10
IR DM45E drill	0.20–0.40	20.3–28.7	0.05	2.80	0.02
DrillTech DK40 drill	0.07–0.12	12.0–15.7	0.07	6.25	0.01

[1]Wind velocity equivalent = (4000 $\sqrt{\Delta p_{cab}}$) ft/min × 0.011364 miles · min/ft · hr @ STP.

During these retrofits, any reasonably repairable cab enclosure cracks, gaps, or openings were sealed with silicon and closed-cell foam tape. The cabs had varying degrees of enclosure integrity, indicated by their differences in cab pressures. A wind velocity equivalent for the measured cab pressures was also calculated using the velocity pressure relationship of the general orifice equation (Equation E-15), assuming air density at standard temperature and pressure (STP, 70 °F and 29.92 in Hg). These wind velocity equivalents are also shown in Table 6 and generally indicate the wind velocity resistance of the cab. Field evaluation of these cab systems were conducted with personal gravimetric respirable samplers during three to seven operating shifts.

These field study results show that all of the cab filtration systems reduced outside dust penetration into the enclosure, but suggest that enclosed cab integrity was a factor for their range of penetration performance. Lower respirable dust cab penetration was observed for two tighter cabs that operated from 0.07 to 0.40 in w.g., or wind velocity equivalents of 12 to 28.7 mph, respectively. The three cabs that operated from 0 to 0.015 in w.g. or wind velocity equivalents of 0 to 5.6 mph, respectively, had higher respirable dust penetrations. Since these three cabs had achieved cab pressures ≤0.015 in w.g. with 70 ft^3/min or more of intake airflow, one can infer that these cabs had significantly larger leakage areas than the laboratory cab test stand. These leakage areas comprised tough-to-seal enclosure openings around movable mechanical control linkages, behind control panels and from other unidentifiable openings on the cab, which are susceptible to penetrating air velocities by wind or the equipment itself such as engine fans and/or tire movement. Also, the Euclid R-50 truck could easily exceed the wind velocity equivalent of the cab pressure during its traveling operation. Although all of these field studies showed that enclosed cab filtration systems reduced outside dust penetration, they also imply that better enclosure integrity ensures increased pressurization and resistance to outside penetration from wind or high air velocity sources during field operations.

CONCLUSIONS

Cab air filtration system factors were experimentally studied in the laboratory for submicron particulate penetration into the cab enclosure. Both series of experiments indicated that the intake filter efficiency and recirculation filter were the two most influential factors on cab penetration. The higher-efficiency intake filter (>99% capture efficiency) changed the cab penetration by an order of magnitude over the lower-efficiency intake filter (from 29% to 44% capture efficiency). Using a recirculation filter (72.4% capture efficiency) further reduced cab penetration, usually by an order of magnitude over the intake air filter alone. The recirculation filter also significantly decreased the decay time needed for the cab interior concentrations to go down and stabilize after the cab door was closed. The average decay times ranged from 16 to 29 min without the recirculation filter and from 6 to 11 min with the recirculation filter. Thus, a recirculation filter mutually reduced cab penetration and exposure time to higher peak concentrations after the cab door was closed.

Air leakage around the intake filter was another significant factor on cab *Pen* and was directly related to the pressure differential across the leak. Loading and leakage interactions with the intake filter efficiency were also found to be statistically significant with cab *Pen*. Wind had the least impact on cab *Pen* between the calm and 10-mph wind velocities tested and was only found to be significant as an interaction with intake filter loading without the intake pressurizer fan.

Adding an intake pressurizer fan to the cab filtration system increased intake airflow and cab pressure significantly with negligible changes to recirculation airflow and only small changes to cab *Pen*. The lower-efficiency intake filter showed decreased capture efficiency at higher intake airflow rates, slightly increasing cab penetration with the pressurizer. The higher-efficiency intake filter showed negligible changes in filter efficiency and cab penetration at higher intake airflows with the pressurizer. Higher intake airflows from the pressurizer increased the negative differential pressure across the intake filter and increased the positive differential pressure inside the cab. Although cab pressure was directly related to intake air quantity, it does not necessarily reflect the intake air quality and overall cab penetration.

Regression analyses of the laboratory cab test stand results corroborated the significance of these experimental test factors on cab *Pen*, but a more general mathematical penetration model was formulated with respect to the cab's filtration system operating variables. It models cab *Pen* in terms of intake filter efficiency, intake air quantity, intake air leakage, recirculation filter efficiency, recirculation filter quantity, and outside wind infiltration. This mathematical model was validated by the experimental test data and can be used to assess cab filtration penetration based on these cab filtration design variables.

ACKNOWLEDGMENT

The authors express their appreciation to Thomas Mal, Engineering Technician, NIOSH-PRL, for his assistance with the setup and execution of these experiments.

REFERENCES

ACGIH [1973]. Threshold limit values for chemical substances in workroom air adopted by the ACGIH for 1973. Cincinnati, OH: American Conference of Governmental Industrial Hygienists.

Cecala AB, Organiscak JA, Heitbrink WA, Zimmer JA, Fisher T, Gresh RE, Ashley JD [2003]. Reducing enclosed cab drill operator's respirable dust exposure at a surface coal operation using a retrofitted filtration and pressurization system. In: Yernberg WR, ed. Transactions of Society for Mining, Metallurgy, and Exploration, Inc. Vol. 314. Littleton, CO: Society for Mining, Metallurgy, and Exploration, Inc., pp. 31–36.

Cecala AB, Organiscak JA, Zimmer JA, Heitbrink WA, Moyer ES, Schmitz M, Ahrenholtz E, Coppock CC, Andrews EH [2005]. Reducing enclosed cab drill operator's respirable dust exposure with effective filtration and pressurization techniques. J Occup Environ Hyg $2(1)$:54–63.

CFR. Code of Federal regulations. Washington, DC: U.S. Government Printing Office, Office of the Federal Register.

Chekan GJ, Colinet JF [2003]. Retrofit options for better dust control. Aggregates Manag $8(9)$:9–12.

Hartman HL [1961]. Mine ventilation and air conditioning. New York: John Wiley & Sons, pp. 37–38.

Heitbrink WA, Hall RM, Reed LD, Gibbons D [1998]. Review of ambient aerosol test procedures in ASAE standard S525. J Agric Saf Health $4(4)$:255–266.

Heitbrink WA, Thimons ED, Organiscak JA, Cecala AB, Schmitz M, Ahrenholtz E [2000]. Static pressure requirements for ventilated enclosures. In: Proceedings of the Sixth International Symposium on Ventilation for Contaminant Control (Helsinki, Finland, June 4–7, 2000), pp. 97–99.

Myers RH, Montgomery DC [1995]. Response surface methodology: process and product optimization using designed experiments. New York: John Wiley & Sons, Inc.

NIOSH [2003]. Work-related lung disease surveillance report, 2002. Morgantown, WV: U.S. Department of Health and Human Services, Centers for Disease Control and Prevention, National Institute for Occupational Safety and Health, Division of Respiratory Disease Studies, DHHS (NIOSH) Publication No. 2003–111.

NIOSH [2008]. Reducing silica exposures in mining through improved control technologies. [http://www.cdc.gov/niosh/nas/mining/strategicoutcome2.htm]. Date accessed: August 2008.

Organiscak JA, Page SJ [1999]. Field assessment of control techniques and long-term dust variability for surface coal mine rock drills and bulldozers. Int J Surf Min Reclam Env 13:165–172.

Organiscak JA, Cecala AB, Thimons ED, Heitbrink WA, Schmitz M, Ahrenholtz E [2003]. NIOSH/industry collaborative efforts show improved mining equipment cab dust protection. In: Yernberg WR, ed. Transactions of Society for Mining, Metallurgy, and

Exploration, Inc. Vol. 314. Littleton, CO: Society for Mining, Metallurgy, and Exploration, Inc., pp. 145–152.

Parobeck PS, Tomb TF [2000]. MSHA's programs to quantify the crystalline silica content of respirable mine dust samples. SME preprint 00-159. Littleton, CO: Society for Mining, Metallurgy, and Exploration, Inc.

Streeter VL, Wylie EB [1979]. Fluid mechanics. 7th ed. New York: McGraw-Hill Book Co., pp. 342–349.

Tomb TF, Gero AJ, Kogut J [1995]. Analysis of quartz exposure data obtained from underground and surface coal mining operations. Appl Occup Environ Hyg *10*(12):1019–1026.

APPENDIX A.—HALF-FRACTION EXPERIMENTAL DESIGN

Table A-1 shows the construction of two experimental half-fractions. The lower intake filter efficiency, unloaded intake filter, no intake leak, no recirculation filter, and no wind factor levels were coded with −1. The higher intake filter efficiency (*a*), loaded intake filter (*b*), intake leak (*c*), recirculation filter (*d*), and wind (*e*) factor levels were coded with +1. Each half-fraction has a full two-level factorial design for the four cab factor configurations (*ABCD*), with the plus and minus wind factor (*E*) levels in each half-fraction identified by the plus and minus sign of the highest-order interaction (identity (*I*) = +*ABCDE* and −*ABCDE*). This partitioned the experimental runs into two identity test blocks with the highest-order interaction (*ABCDE*) confounded [Myers and Montgomery 1995].

Table A-1.—Half-fraction experimental design

| Run | Experimental condition | Factor level | | | | Block 1 $I = +ABCDE$ | | Block 2 $I = -ABCDE$ | |
		A Intake filter efficiency	B Intake filter loading	C Intake air leakage	D Recirculation filter	E Wind	Treatment Combination	E Wind	Treatment Combination
1	*1*	−1	−1	−1	−1	+1	e	−1	*1*
2	*a*	+1	−1	−1	−1	−1	a	+1	ae
3	*b*	−1	+1	−1	−1	−1	b	+1	be
4	*ab*	+1	+1	−1	−1	+1	abe	−1	ab
5	*c*	−1	−1	+1	−1	−1	c	+1	ce
6	*ac*	+1	−1	+1	−1	+1	ace	−1	ac
7	*bc*	−1	+1	+1	−1	+1	bce	−1	bc
8	*abc*	+1	+1	+1	−1	−1	abc	+1	abce
9	*d*	−1	−1	−1	+1	−1	d	+1	de
10	*ad*	+1	−1	−1	+1	+1	ade	−1	ad
11	*bd*	−1	+1	−1	+1	+1	bde	−1	bd
12	*abd*	+1	+1	−1	+1	−1	abd	+1	abde
13	*cd*	−1	−1	+1	+1	+1	cde	−1	cd
14	*acd*	+1	−1	+1	+1	−1	acd	+1	acde
15	*bcd*	−1	+1	+1	+1	−1	bcd	+1	bcde
16	*abcd*	+1	+1	+1	+1	+1	abcde	−1	abcd

APPENDIX B.—EXPERIMENTAL TEST DATA

Table B-1.—First half-fraction of test data without intake pressurizer ($I = +ABCDE$)

(Tests sorted by run number or experimental condition)

Test and period Nos.	Run No.	Test condition	Cab operating parameters					Wind v_w ft/min	Wet-bulb temp., °F	Dry-bulb temp., °F	Barometric press., in Hg	Decay time, min	Start [‡]C_1 counts/L	Last 15-min test average			Test average [‡]C_3 counts/L	Cab Pen C_1/C_3
			Q_I ft³/min	$-\Delta p_I$ in w.g.	[†]Q_L L/min	$+\Delta p_c$ in w.g.	Q_R ft³/min							[‡]C_1 counts/L	[‡]C_2 counts/L	[‡]C_3 counts/L		
T2.P1	1	e	47.9	0.14	0.3	0.28	355	877	44.2	56.5	29.14	1	36,350	30,383	NA	48,760	46,027	0.623
T2.P2	1	e	48.9	0.14	0.3	0.28	355	904	42.0	53.5	29.14	4	49,244	33,545	NA	52,633	51,773	0.637
T18.P1	1	e	45.4	0.15	0.3	0.26	368	878	54.5	66.8	29.09	2	19,738	14,126	15,231	22,916	21,432	0.616
T18.P2	1	e	46.0	0.15	0.3	0.26	362	886	53.5	65.8	29.09	2	23,008	18,485	19,371	29,287	26,930	0.631
T35.P1	1	e	46.8	0.15	0.3	0.25	360	883	44.5	50.0	28.98	37	26,179	10,791	11,479	17,224	20,929	0.627
T35.P2	1	e	48.0	0.15	0.2	0.25	358	876	41.5	48.5	29.00	2	12,758	10,367	11,622	17,534	14,098	0.591
T27.P1	2	a	23.0	0.51	0.4	0.07	378	Calm	53.5	63.8	29.03	32	68,013	2,067	117	75,368	76,331	0.027
T27.P2	2	a	23.0	0.51	0.5	0.07	385	Calm	52.5	64.2	29.01	25	55,231	1,656	67	45,682	48,045	0.036
T28.P1	2	a	22.7	0.51	0.4	0.07	385	Calm	54.0	65.8	28.93	34	38,838	1,597	64	42,482	49,953	0.038
T28.P2	2	a	22.7	0.51	0.4	0.07	388	Calm	53.5	66.0	28.92	27	34,351	1,747	59	38,713	38,641	0.045
T6.P1	3	b	20.5	0.53	0.4	0.07	368	Calm	52.0	66.5	28.98	37	23,502	10,710	11,433	19,563	21,699	0.547
T6.P2	3	b	20.6	0.53	0.4	0.07	378	Calm	51.5	66.5	29.00	13	17,700	10,008	10,453	18,272	17,992	0.548
T12.P1	3	b	20.8	0.52	0.4	0.06	370	Calm	60.0	74.8	28.96	10	28,949	16,920	17,815	31,492	31,050	0.537
T12.P2	3	b	21.0	0.52	0.4	0.06	372	Calm	60.0	75.0	28.94	38	30,752	17,653	18,363	32,866	33,226	0.537
T19.P1	4	abe	14.1	0.52	0.3	0.09	385	888	47.5	56.8	29.26	15	108,789	19,914	374	140,665	133,737	0.142
T19.P2	4	abe	14.4	0.52	0.3	0.09	382	878	46.5	57.2	29.24	39	124,576	15,767	287	113,173	120,995	0.139
T20.P1	4	abe	14.0	0.52	0.3	0.08	382	881	49.8	65.0	29.16	36	61,142	6,452	111	53,636	57,970	0.120
T20.P2	4	abe	13.8	0.52	0.3	0.08	385	869	49.0	64.0	29.14	38	45,947	5,456	96	45,597	46,830	0.120
T33.P1	4	abe	14.8	0.52	0.3	0.09	390	876	47.0	56.0	29.26	19	68,911	9,846	153	76,538	71,388	0.129
T33.P2	4	abe	14.6	0.53	0.3	0.08	390	894	44.8	55.5	29.24	39	62,273	7,191	111	56,837	59,971	0.127
T9.P1	5	c	49.6	0.16	24.5	0.25	360	Calm	55.5	73.2	29.33	32	21,208	14,829	14,951	22,763	23,046	0.651
T9.P2	5	c	49.8	0.16	24.6	0.26	360	Calm	55.0	74.0	29.32	16	20,580	14,638	14,726	22,350	22,347	0.655
T10.P1	5	c	49.8	0.16	24.3	0.26	362	Calm	56.5	76.5	29.26	27	20,229	13,777	13,772	21,444	21,786	0.642
T10.P2	5	c	50.0	0.16	24.3	0.26	362	Calm	56.0	76.5	29.26	6	19,198	14,024	13,866	21,509	21,570	0.652
T8.P1	6	ace	23.5	0.45	44.1	0.12	375	913	41.5	52.8	29.28	24	19,999	2,215	267	24,905	24,468	0.089
T8.P2	6	ace	24.0	0.45	44.5	0.12	370	900	38.5	47.0	29.26	20	22,644	2,111	268	24,047	24,318	0.088

NA Not available.
[†]The mass flowmeter analog output had a several tenths of flow bias at 0.0 L/min on the display.
[‡]The particle counter concentrations are for particle diameter sizes ranging from 0.3 to 1.0 μm.

Table B-1.—First half-fraction of test data without intake pressurizer ($I = +ABCDE$) —Continued

(Tests sorted by run number or experimental condition)

Test and period Nos.	Run No.	Test condition	Cab operating parameters						Wind v_w ft/min	Wet-bulb temp., °F	Dry-bulb temp., °F	Barometric press., in Hg	Decay time, min	Start $^‡C_1$ counts/L	Last 15-min test average				Test average $^‡C_3$ counts/L	Cab Pen C_1/C_3
			Q_I ft³/min	$-\Delta p_f$ in w.g.	$^†Q_L$ L/min	$+\Delta p_c$ in w.g.	Q_R ft³/min								$^‡C_1$ counts/L	$^‡C_2$ counts/L	$^‡C_3$ counts/L		
T16.P1	6	ace	21.6	0.45	42.9	0.10	385		900	53.8	69.0	28.60	16	20,820	2,671	329	20,187	21,033	0.132
T16.P2	6	ace	21.6	0.45	42.8	0.10	385		901	53.5	69.0	28.58	26	16,619	1,760	285	17,849	17,898	0.099
T3.P1	7	bce	22.2	0.46	44.5	0.12	390		882	46.0	61.0	28.91	5	25,426	17,595	19,271	31,207	29,124	0.564
T3.P2	7	bce	22.3	0.46	44.7	0.12	390		871	46.0	59.0	28.86	4	35,720	26,094	27,982	43,519	43,433	0.600
T15.P1	7	bce	21.9	0.46	43.5	0.10	388		881	54.8	66.2	28.66	24	42,535	25,569	27,777	45,624	46,160	0.560
T15.P2	7	bce	22.2	0.46	43.5	0.10	390		902	53.8	66.2	28.64	28	43,403	24,521	25,326	41,143	42,649	0.596
T4.P1	8	abc	16.2	0.56	50.4	0.05	385		Calm	41.5	52.8	29.28	19	20,064	2,632	1,803	19,666	19,697	0.134
T4.P2	8	abc	16.2	0.56	50.2	0.05	385		Calm	38.5	47.0	29.26	15	17,437	3,005	1,810	19,318	19,174	0.156
T31.P1	8	abc	15.5	0.56	50.7	0.04	390		Calm	46.0	55.8	28.68	34	51,404	4,534	2,660	29,908	37,297	0.152
T31.P2	8	abc	15.6	0.56	50.8	0.04	388		Calm	43.2	54.2	28.68	37	33,377	9,734	6,072	54,894	70,390	0.177
T23.P1	9	d	57.8	0.22	0.5	0.32	322		Calm	60.8	74.5	28.98	13	18,256	6,693	34,396	51,091	51,358	0.131
T23.P2	9	d	57.7	0.22	0.4	0.33	322		Calm	61.0	75.2	28.98	8	20,983	6,445	32,841	48,277	48,340	0.134
T32.P1	9	d	59.5	0.22	0.4	0.31	315		Calm	42.8	51.8	28.64	5	9,415	3,734	20,757	30,552	30,288	0.122
T32.P2	9	d	58.4	0.22	0.4	0.32	315		Calm	41.8	51.8	28.65	12	9,656	3,219	17,478	26,358	26,968	0.122
T13.P1	10	ade	26.8	0.62	0.3	0.12	332		881	56.5	76.5	29.26	3	17,792	127	42	56,133	54,626	0.002
T13.P2	10	ade	27.1	0.63	0.3	0.12	338		881	56.0	76.5	29.26	13	21,105	114	43	55,910	56,304	0.002
T24.P1	10	ade	26.2	0.62	0.4	0.12	340		878	60.2	70.8	28.86	4	18,351	133	50	47,345	50,721	0.003
T24.P2	10	ade	26.5	0.62	0.3	0.12	345		877	59.2	69.5	28.84	2	14,981	135	42	43,717	43,800	0.003
T11.P1	11	bde	23.7	0.67	0.4	0.10	335		880	55.2	66.5	29.01	2	19,025	2,363	28,462	47,442	46,644	0.050
T11.P2	11	bde	24.0	0.67	0.3	0.10	335		888	54.8	66.0	29.00	2	17,402	2,366	27,955	46,420	46,796	0.051
T30.P1	11	bde	23.6	0.67	0.3	0.10	338		886	62.0	66.2	28.67	1	17,662	2,836	34,800	62,857	62,629	0.045
T30.P2	11	bde	23.9	0.67	0.3	0.10	338		894	62.2	66.5	28.65	10	26,015	2,251	26,580	47,826	50,150	0.047
T14.P1	12	abd	17.3	0.76	0.4	0.04	348		Calm	62.0	75.8	28.82	18	8,568	93	32	25,012	25,509	0.004
T14.P2	12	abd	17.2	0.76	0.3	0.05	350		Calm	61.0	76.0	28.78	5	7,496	116	29	23,280	23,499	0.005
T17.P1	12	abd	17.8	0.76	0.5	0.04	345		Calm	63.5	78.0	29.06	8	15,870	166	71	48,640	50,329	0.003
T17.P2	12	abd	17.8	0.76	0.6	0.05	348		Calm	62.0	78.0	29.08	13	11,806	158	54	37,561	39,369	0.004
T5.P1	13	cde	56.0	0.19	25.2	0.37	325		904	46.0	55.8	28.92	1	25,003	7,964	43,622	60,761	59,146	0.131
T5.P2	13	cde	56.8	0.19	25.5	0.37	320		920	43.2	54.8	28.92	1	32,682	9,475	50,781	69,687	68,893	0.136

†The mass flowmeter analog output had a several tenths of flow bias at 0.0 L/min on the display.
‡The particle counter concentrations are for particle diameter sizes ranging from 0.3 to 1.0 μm.

Table B-1.—First half-fraction of test data without intake pressurizer ($I = +ABCDE$) —Continued

(Tests sorted by run number or experimental condition)

Test and period Nos.	Run No.	Test condition	Cab operating parameters					Wind v_w ft/min	Wet-bulb temp., °F	Dry-bulb temp., °F	Baro-metric press., in Hg	Decay time, min	Start ‡C_1 counts/L	Last 15-min test average			Test average ‡C_3 counts/L	Cab Pen C_1/C_3
			Q_I ft³/min	$-\Delta p_f$ in w.g.	$^{\dagger}Q_L$ L/min	$+\Delta p_c$ in w.g.	Q_R ft³/min							‡C_1 counts/L	‡C_2 counts/L	‡C_3 counts/L		
T26.P1	13	cde	57.8	0.21	26.2	0.33	322	891	51.0	64.5	29.10	14	19,143	4,789	25,547	39,018	40,073	0.123
T26.P2	13	cde	58.1	0.21	26.2	0.33	320	908	51.2	65.5	29.10	21	14,223	4,308	22,122	32,998	33,961	0.131
T7.P1	14	acd	30.1	0.65	55.2	0.11	335	Calm	51.5	66.8	29.30	5	4,792	153	168	14,782	14,327	0.010
T7.P2	14	acd	30.2	0.65	55.3	0.11	338	Calm	50.5	67.0	29.32	2	4,966	176	169	14,713	14,341	0.012
T25.P1	14	acd	30.0	0.64	54.8	0.10	332	Calm	53.5	66.0	29.20	3	25,024	849	1,203	96,535	92,181	0.009
T25.P2	14	acd	29.6	0.64	54.9	0.10	335	Calm	53.0	67.8	29.19	20	27,662	648	800	70,240	75,281	0.009
T1.P1	15	bcd	26.5	0.71	57.0	0.09	340	Calm	54.2	71.2	29.16	1	7,037	1,127	12,350	20,027	19,791	0.056
T1.P2	15	bcd	26.6	0.70	56.8	0.09	338	Calm	53.0	71.8	29.16	4	8,257	1,014	11,057	18,154	18,127	0.056
T29.P1	15	bcd	25.6	0.69	56.4	0.08	338	Calm	62.0	67.5	28.77	2	13,415	2,218	24,288	42,218	40,740	0.053
T29.P2	15	bcd	25.9	0.69	56.4	0.08	332	Calm	62.0	68.0	28.76	1	14,591	2,840	31,911	54,915	49,702	0.052
T36.P1	15	bcd	27.2	0.70	58.0	0.08	338	Calm	49.8	61.2	29.05	23	19,350	2,094	21,870	35,137	41,814	0.060
T36.P2	15	bcd	27.0	0.69	57.9	0.08	335	Calm	49.5	62.0	29.06	21	7,295	878	9,095	15,107	17,747	0.058
T21.P1	16	abcde	18.6	0.72	56.7	0.09	348	882	51.5	62.5	29.20	2	38,797	1,080	9,515	110,525	104,355	0.010
T21.P2	16	abcde	19.2	0.72	57.0	0.09	345	883	49.5	60.5	29.20	2	36,589	1,167	10,236	114,693	112,835	0.010
T22.P1	16	abcde	18.8	0.71	56.1	0.09	345	884	51.2	64.0	29.14	16	26,952	601	5,032	63,261	65,637	0.010
T22.P2	16	abcde	18.9	0.72	56.3	0.08	348	854	49.2	61.5	29.14	2	18,984	664	5,334	66,037	62,081	0.010
T34.P1	16	abcde	20.2	0.72	57.0	0.09	332	883	43.8	56.2	29.16	23	16,087	541	4,930	52,772	58,905	0.010
T34.P2	16	abcde	20.1	0.72	57.6	0.09	338	878	43.0	55.5	29.14	4	8,405	288	2,172	23,865	23,568	0.012

†The mass flowmeter analog output had a several tenths of flow bias at 0.0 L/min on the display.
‡The particle counter concentrations are for particle diameter sizes ranging from 0.3 to 1.0 µm.

Table B-2.—Second half-fraction of test data without intake pressurizer ($I = -ABCDE$)

(Tests sorted by run number or experimental condition)

Test and period Nos.	Run No.	Test condition	Cab operating parameters					Wind v_w ft/min	Wet-bulb temp., °F	Dry-bulb temp., °F	Barometric press., in Hg	Decay time, min	Start ‡C_1 counts/L	Last 15-min test average				Test average ‡C_3 counts/L	Cab Pen C_1/C_3
			Q_I ft³/min	$-\Delta p_f$ in w.g.	†Q_L L/min	$+\Delta p_c$ in w.g.	Q_R ft³/min							‡C_1 counts/L	‡C_2 counts/L	‡C_3 Counts/L		
T56.P1	1	1	50.4	0.18	0.4	0.21	355	Calm	52.0	69.8	29.01	24	20,155	14,089	13,331	20,781	21,032	0.678
T56.P2	1	1	50.2	0.17	0.3	0.21	355	Calm	52.0	70.0	29.00	32	20,189	13,375	12,530	19,371	19,697	0.690
T69.P1	1	1	48.7	0.17	0.4	0.21	352	Calm	57.0	74.5	28.39	2	12,343	8,007	7,934	14,026	12,923	0.571
T69.P2	1	1	48.2	0.17	0.4	0.21	358	Calm	57.0	75.0	28.38	6	13,412	8,158	7,978	14,660	14,898	0.556
T75.P1	1	1	49.0	0.17	0.4	0.23	368	Calm	63.2	78.5	28.96	36	97,913	58,335	58,500	87,912	100,363	0.664
T75.P2	1	1	49.2	0.17	0.4	0.23	368	Calm	62.8	79.5	28.96	34	74,747	49,686	49,713	74,271	76,089	0.669
T41.P1	2	ae	21.3	0.46	0.3	0.10	390	872	41.8	49.5	28.71	34	23,484	1,239	34	28,772	28,632	0.043
T41.P2	2	ae	22.2	0.47	0.3	0.10	390	883	39.0	47.8	28.70	33	26,731	1,213	37	30,079	29,987	0.040
T47.P1	2	ae	21.0	0.47	0.3	0.08	385	862	47.8	50.8	28.52	30	147,368	2,660	116	96,816	113,274	0.027
T47.P2	2	ae	21.8	0.46	0.2	0.10	388	850	42.5	50.0	28.50	36	83,022	3,211	98	81,250	80,983	0.040
T40.P1	3	be	20.8	0.48	0.3	0.10	368	885	41.8	52.0	28.76	31	19,510	8,834	9,384	15,889	16,900	0.556
T40.P2	3	be	21.2	0.49	0.2	0.09	375	873	39.8	49.8	28.77	3	15,726	11,553	12,699	27,120	23,105	0.426
T61.P1	3	be	20.6	0.49	0.3	0.08	378	850	44.0	52.0	29.10	28	10,706	6,863	6,347	11,233	11,056	0.611
T61.P2	3	be	20.7	0.49	0.3	0.08	375	854	42.0	52.0	29.10	8	11,979	7,605	7,616	13,140	12,707	0.579
T45.P1	4	ab	14.8	0.57	0.4	0.03	398	Calm	50.5	66.5	28.77	37	24,612	1,817	27	26,876	26,569	0.068
T45.P2	4	ab	14.7	0.56	0.4	0.03	395	Calm	48.5	64.0	28.77	26	42,633	2,280	NA	36,394	41,872	0.063
T48.P1	4	ab	14.6	0.56	0.5	0.03	390	Calm	52.2	66.0	28.48	37	48,610	2,138	65	51,517	52,842	0.042
T48.P2	4	ab	14.5	0.56	0.4	0.03	392	Calm	53.5	68.0	28.48	31	45,221	1,857	53	45,952	47,447	0.040
T70.P1	4	ab	14.0	0.56	0.4	0.03	398	Calm	56.5	76.0	28.35	30	14,366	1,725	26	12,324	12,829	0.140
T70.P2	4	ab	14.0	0.56	0.3	0.03	392	Calm	57.5	77.0	28.36	30	11,637	2,140	29	15,365	12,563	0.139
T73.P1	4	ab	15.2	0.58	0.4	0.03	365	Calm	54.5	71.0	29.34	25	14,967	2,160	36	11,924	12,184	0.181
T73.P2	4	ab	15.2	0.58	0.4	0.03	375	Calm	53.5	72.0	29.32	37	12,375	2,007	38	12,323	12,127	0.163
T74.P1	4	ab	15.1	0.57	0.5	0.03	388	Calm	60.8	75.0	29.02	29	113,185	6,379	263	121,196	120,629	0.053
T74.P2	4	ab	14.7	0.57	0.6	0.03	390	Calm	61.2	77.0	29.02	22	107,085	8,642	257	116,388	116,036	0.074
T39.P1	5	ce	47.2	0.14	21.9	0.25	355	870	44.2	49.0	28.79	2	93,959	73,832	81,208	116,720	113,966	0.633
T39.P2	5	ce	47.4	0.14	21.8	0.25	338	886	42.2	52.2	28.78	38	110,092	42,963	44,266	63,351	86,921	0.678

NA Not available.

†The mass flowmeter analog output had a several tenths of flow bias at 0.0 L/min on the display.
‡The particle counter concentrations are for particle diameter sizes ranging from 0.3 to 1.0 μm.

Table B-2.—Second half-fraction of test data without intake pressurizer ($I = -ABCDE$)—Continued

(Tests sorted by run number or experimental condition)

Test and period Nos.	Run No.	Test condition	Cab operating parameters						Wind v_w ft/min	Wet-bulb temp., °F	Dry-bulb temp., °F	Barometric press., in Hg	Decay time, min	Start ‡C_1 counts/L	Last 15-min test average			Test average ‡C_3 counts/L	Cab Pen C_1/C_3
			Q_I ft³/min	$-\Delta p_f$ in w.g.	†Q_L L/min	$+\Delta p_c$ in w.g.	Q_R ft³/min								‡C_1 counts/L	‡C_2 counts/L	‡C_3 counts/L		
T57.P1	5	ce	49.8	0.15	22.8	0.24	368		864	44.5	54.5	29.10	32	39,991	27,509	27,724	42,126	42,538	0.653
T57.P2	5	ce	50.6	0.15	22.9	0.24	355		860	41.5	53.0	29.08	6	40,984	28,776	28,960	43,224	42,182	0.666
T71.P1	5	ce	49.8	0.16	22.6	0.21	348		934	46.0	56.5	28.83	4	36,662	24,783	27,100	41,715	40,991	0.594
T71.P2	5	ce	50.6	0.16	22.5	0.21	352		941	45.2	55.2	28.83	4	39,445	25,827	27,796	41,850	40,911	0.617
T37.P1	6	ac	23.6	0.49	46.7	0.07	378		Calm	57.8	73.2	28.77	34	43,314	5,168	603	47,992	48,887	0.108
T37.P2	6	ac	23.3	0.49	46.6	0.07	382		Calm	57.2	73.2	28.76	16	41,007	5,065	579	43,561	44,625	0.116
T50.P1	6	ac	25.0	0.50	47.7	0.06	388		Calm	48.8	65.0	29.10	36	36,721	3,271	216	32,015	34,194	0.102
T50.P2	6	ac	24.8	0.50	47.4	0.06	380		Calm	49.2	66.5	29.08	15	27,840	3,253	215	27,905	28,741	0.117
T49.P1	7	bc	22.2	0.53	48.9	0.05	382		Calm	49.5	64.5	29.18	36	25,911	12,630	11,987	21,148	23,317	0.597
T49.P2	7	bc	22.2	0.53	48.8	0.05	382		Calm	48.0	64.2	29.17	6	22,281	13,960	12,892	22,503	23,377	0.620
T63.P1	7	bc	22.3	0.51	47.3	0.06	368		Calm	56.5	73.0	28.67	8	16,412	9,860	9,366	16,898	16,362	0.584
T63.P2	7	bc	22.0	0.51	47.3	0.06	370		Calm	56.5	73.5	28.67	6	17,996	11,855	10,868	18,622	17,977	0.637
T42.P1	8	abce	15.6	0.51	46.9	0.08	395		887	42.5	50.5	28.67	37	25,057	5,216	2,748	27,919	28,724	0.187
T42.P2	8	abce	15.9	0.51	47.2	0.08	390		881	41.2	48.8	28.68	36	23,150	4,381	2,150	20,745	22,342	0.211
T46.P1	8	abce	14.6	0.50	46.6	0.09	395		896	45.5	56.0	28.70	26	21,655	4,891	1,703	25,676	24,496	0.190
T46.P2	8	abce	15.6	0.50	46.8	0.09	392		903	46.0	54.5	28.68	12	26,044	6,610	2,390	35,440	32,832	0.187
T62.P1	9	de	60.4	0.23	0.3	0.29	318		854	43.0	54.0	28.68	1	6,325	2,389	12,131	18,113	17,617	0.132
T62.P2	9	de	61.0	0.23	0.2	0.28	318		859	41.5	52.5	29.08	15	8,753	2,404	11,933	17,666	17,864	0.136
T68.P1	9	de	57.2	0.22	0.3	0.29	300		916	52.0	66.0	28.64	2	37,260	12,558	60,775	87,910	87,437	0.143
T68.P2	9	de	57.5	0.22	0.3	0.29	305		927	52.0	66.5	28.60	1	39,978	13,482	63,812	90,801	90,380	0.148
T51.P1	10	ad	29.3	0.67	0.4	0.08	330		Calm	50.5	65.5	29.00	14	5,942	137	16	24,450	24,993	0.006
T51.P2	10	ad	29.4	0.67	0.4	0.08	328		Calm	49.5	66.5	29.02	2	5,136	136	17	25,561	25,307	0.005
T59.P1	10	ad	28.0	0.67	0.5	0.07	325		Calm	58.8	74.0	28.82	9	17,526	213	48	61,746	62,056	0.003
T59.P2	10	ad	28.0	0.67	0.7	0.08	330		Calm	59.5	75.0	28.83	11	17,859	221	46	60,687	61,847	0.004
T38.P1	11	bd	23.6	0.70	0.3	0.07	338		Calm	56.2	73.0	28.74	13	10,277	1,949	21,522	36,452	37,221	0.053
T38.P2	11	bd	23.6	0.70	0.4	0.07	335		Calm	56.0	73.0	28.72	17	9,805	1,758	19,451	32,804	33,228	0.054
T52.P1	11	bd	24.8	0.72	0.4	0.06	335		Calm	49.5	65.5	29.00	7	7,104	1,287	14,020	24,389	24,458	0.053
T52.P2	11	bd	24.8	0.72	0.4	0.06	335		Calm	49.0	65.5	28.98	23	8,381	1,396	15,388	26,747	28,744	0.052

†The mass flowmeter analog output had a several tenths of flow bias at 0.0 L/min on the display.
‡The particle counter concentrations are for particle diameter sizes ranging from 0.3 to 1.0 μm.

Table B-2.—Second half-fraction of test data without intake pressurizer ($I = -ABCDE$) —Continued

(Tests sorted by run number or experimental condition)

Test and period Nos.	Run No.	Test condition	Cab operating parameters					Wind v_w ft/min	Wet-bulb temp., °F	Dry-bulb temp., °F	Barometric press., in Hg	Decay time, min	Start ‡C_1 counts/L	Last 15-min test average			Test average ‡C_3 counts/L	Cab Pen C_1/C_3
			Q_I ft³/min	$-\Delta p_f$ in w.g.	†Q_L L/min	$+\Delta p_c$ in w.g.	Q_R ft³/min							‡C_1 counts/L	‡C_2 counts/L	‡C_3 counts/L		
T65.P1	12	*abde*	17.4	0.72	0.3	0.07	348	840	47.5	58.0	28.80	6	26,445	507	110	77,348	78,349	0.007
T65.P2	12	*abde*	17.7	0.73	0.3	0.07	348	834	45.5	57.0	28.80	5	26,385	435	97	66,400	67,559	0.007
T66.P1	12	*abde*	17.4	0.72	0.3	0.07	345	854	48.0	61.0	28.82	6	17,227	290	64	46,709	46,849	0.006
T66.P2	12	*abde*	17.6	0.73	0.3	0.07	348	851	46.8	59.8	28.82	10	16,139	284	62	45,013	45,445	0.006
T43.P1	13	*cd*	61.0	0.23	30.0	0.30	328	Calm	50.0	65.0	29.04	4	7,667	3,266	15,842	24,051	23,767	0.136
T43.P2	13	*cd*	60.6	0.23	30.1	0.30	325	Calm	49.5	65.8	29.04	9	7,070	3,018	14,694	21,744	21,869	0.139
T54.P1	13	*cd*	59.8	0.23	29.9	0.29	315	Calm	50.0	68.0	29.03	1	8,148	3,344	15,621	24,063	23,373	0.139
T54.P2	13	*cd*	59.2	0.23	29.9	0.28	312	Calm	50.5	68.0	29.02	1	8,059	3,413	15,848	23,727	22,943	0.144
T60.P1	14	*acde*	28.9	0.63	52.3	0.10	328	841	52.0	65.0	28.79	2	25,496	754	604	80,490	79,988	0.009
T60.P2	14	*acde*	29.0	0.63	52.3	0.11	328	846	51.0	64.0	28.77	7	26,992	702	536	74,034	74,746	0.009
T64.P1	14	*acde*	29.4	0.63	52.0	0.10	318	852	59.0	60.5	28.64	3	11,560	374	360	37,596	35,760	0.010
T64.P2	14	*acde*	30.0	0.63	52.2	0.11	325	848	47.5	59.0	28.63	4	13,778	437	409	42,793	42,304	0.010
T44.P1	15	*bcde*	25.0	0.68	55.5	0.10	342	881	39.8	50.0	28.99	2	14,111	2,232	25,092	39,348	39,135	0.057
T44.P2	15	*bcde*	25.3	0.69	56.0	0.10	340	885	37.8	47.5	28.98	3	12,294	2,211	24,700	38,523	38,612	0.057
T58.P1	15	*bcde*	26.0	0.67	54.6	0.10	345	840	45.5	58.0	29.02	22	14,119	1,945	20,433	34,134	37,462	0.057
T58.P2	15	*bcde*	26.6	0.67	54.8	0.10	335	856	44.5	57.0	29.00	7	9,735	1,603	17,180	28,102	28,362	0.057
T53.P1	16	*abcd*	19.8	0.76	59.7	0.04	345	Calm	50.8	66.5	29.06	7	6,171	308	1,666	22,615	22,633	0.014
T53.P2	16	*abcd*	19.6	0.76	59.8	0.05	342	Calm	50.2	67.2	29.06	2	5,596	309	1,690	22,302	22,214	0.014
T67.P1	16	*abcd*	20.0	0.74	58.5	0.04	330	Calm	56.0	73.5	28.72	20	21,715	810	7,824	78,492	77,926	0.010
T67.P2	16	*abcd*	19.8	0.74	58.4	0.04	340	Calm	55.5	74.8	28.76	1	21,670	1,261	12,870	92,369	85,440	0.014
T72.P1	16	*abcd*	20.2	0.77	60.3	0.04	342	Calm	52.5	68.0	28.85	17	6,198	224	1,890	21,231	21,628	0.011
T72.P2	16	*abcd*	20.0	0.77	60.0	0.04	338	Calm	52.5	69.5	28.86	3	6,212	232	1,776	19,579	19,751	0.012

†The mass flowmeter analog output had a several tenths of flow bias at 0.0 L/min on the display.
‡The particle counter concentrations are for particle diameter sizes ranging from 0.3 to 1.0 µm.

Table B-3.—Pressurizer test data
(Tests sorted by run number or experimental condition)

Test and period Nos.	Run No.	Test condition	Cab operating parameters					Wind	Wet-bulb temp., °F	Dry-bulb temp., °F	Barometric press., in Hg	Decay time, min	Start	Last 15-min test average			Test average	Cab Pen
			Q_I ft³/min	$-\Delta p_f$ in w.g.	$^\dagger Q_L$ L/min	$+\Delta p_c$ in w.g.	Q_R ft³/min	v_w ft/min					$^\ddagger C_1$ counts/L	$^\ddagger C_1$ counts/L	$^\ddagger C_2$ counts/L	$^\ddagger C_3$ counts/L	$^\ddagger C_3$ counts/L	C_1/C_3
T95.P1	1	1	78.5	0.31	0.4	0.45	340	Calm	57.5	74.5	28.96	17	37,506	29,958	30,474	42,916	43,014	0.698
T95.P2	1	1	78.2	0.31	0.4	0.45	342	Calm	57.0	77.0	28.94	1	39,291	35,225	35,502	48,911	49,714	0.720
T96.P1	1	1	80.2	0.31	0.4	0.45	340	Calm	55.0	71.5	28.79	30	58,901	46,101	46,447	67,065	67,121	0.687
T96.P2	1	1	79.6	0.31	0.4	0.45	342	Calm	54.5	72.5	28.77	32	59,566	44,698	44,452	62,239	63,533	0.718
T85.P1	2	a	38.6	0.88	0.4	0.16	362	Calm	58.0	77.0	29.33	31	9,752	591	14	8,147	8,103	0.073
T85.P2	2	a	38.2	0.88	0.4	0.16	358	Calm	56.0	77.0	29.30	24	7,763	506	14	8,166	8,031	0.062
T87.P1	2	a	38.2	0.88	0.4	0.15	372	Calm	59.5	79.0	29.14	12	17,215	835	34	21,177	20,806	0.039
T87.P2	2	a	38.0	0.88	0.4	0.16	370	Calm	58.0	79.0	29.12	13	17,349	765	41	25,670	24,432	0.030
T78.P1	3	b	29.3	1.01	0.5	0.09	380	Calm	60.0	74.5	29.06	8	45,618	31,271	31,235	52,222	50,169	0.599
T78.P2	3	b	29.2	1.00	0.5	0.10	382	Calm	60.5	75.0	29.04	4	53,277	37,982	38,031	62,221	58,483	0.610
T105.P1	3	b	29.2	0.96	0.4	0.10	382	Calm	58.5	75.0	28.74	36	24,232	13,529	13,799	22,683	24,848	0.596
T105.P2	3	b	29.2	0.96	0.4	0.10	395	Calm	58.5	77.0	28.74	5	21,800	14,739	15,097	24,053	22,459	0.613
T93.P1	4	ab	22.0	1.05	0.4	0.06	382	Calm	53.0	70.5	28.80	14	10,799	1,105	42	9,349	10,029	0.118
T93.P2	4	ab	21.9	1.05	0.4	0.07	382	Calm	51.5	71.0	28.80	16	8,799	1,091	35	8,928	8,879	0.122
T94.P1	4	ab	22.4	1.05	0.5	0.06	385	Calm	55.5	71.5	28.99	31	49,492	2,126	224	57,102	57,170	0.037
T94.P2	4	ab	22.1	1.05	0.4	0.06	380	Calm	55.5	72.0	29.00	20	49,267	2,071	212	56,469	57,726	0.037
T109.P1	4	ab	21.5	1.06	0.6	0.07	392	Calm	61.5	81.0	29.06	21	14,557	1,050	46	14,978	15,067	0.070
T109.P2	4	ab	21.4	1.06	0.5	0.07	392	Calm	61.0	83.0	29.06	21	12,868	857	42	13,987	13,918	0.061
T79.P1	5	c	81.9	0.33	36.1	0.42	348	Calm	64.5	78.5	29.00	34	72,557	55,407	56,148	80,123	81,464	0.692
T79.P2	5	c	82.0	0.32	36.1	0.42	342	Calm	64.0	79.0	28.97	36	67,460	49,438	49,754	70,439	72,598	0.702
T103.P1	5	c	80.2	0.31	35.4	0.44	342	Calm	58.0	73.0	28.74	0	13,130	43,669	42,549	68,716	44,010	0.635
T103.P2	5	c	80.1	0.31	35.4	0.44	342	Calm	60.0	75.0	28.74	28	80,013	48,831	46,637	70,263	75,971	0.695
T92.P1	6	ac	40.7	0.87	64.9	0.14	372	Calm	51.5	67.0	28.77	29	15,167	1,356	1,262	13,313	14,767	0.102
T92.P2	6	ac	40.8	0.84	64.1	0.17	375	Calm	51.5	68.0	28.78	22	12,380	1,191	1,008	11,112	11,908	0.107
T107.P1	6	ac	39.6	0.85	63.4	0.16	370	Calm	60.0	79.0	28.80	34	21,398	1,977	2,082	24,922	25,793	0.079
T107.P2	6	ac	39.6	0.85	63.3	0.16	378	Calm	60.0	79.5	28.81	36	19,245	1,603	1,704	20,150	20,911	0.080

[†] The mass flowmeter analog output had a several tenths of flow bias at 0.0 L/min on the display.
[‡] The particle counter concentrations are for particle diameter sizes ranging from 0.3 to 1.0 µm.

Table B-3.— Pressurizer test data—Continued
(Tests sorted by run number or experimental condition)

Test and period Nos.	Run No.	Test condition	Cab operating parameters					Wind v_w ft/min	Wet-bulb temp., °F	Dry-bulb temp., °F	Barometric press., in Hg	Decay time, min	Start [‡]C_1 counts/L	Last 15-min test average			Test average [‡]C_3 counts/L	Cab Pen C_1/C_3
			Q_I ft³/min	$-\Delta p_F$ in w.g.	$+\Delta p_c$ in w.g.	[†]Q_L L/min	Q_R ft³/min							[‡]C_1 counts/L	[‡]C_2 counts/L	[‡]C_3 counts/L		
T80.P1	7	bc	31.2	0.95	0.11	67.7	370	Calm	64.5	76.5	29.00	3	29,559	23,244	24,477	37,884	35,954	0.614
T80.P2	7	bc	31.4	0.95	0.11	67.5	380	Calm	64.5	77.5	29.00	3	34,375	27,157	28,195	43,807	42,227	0.620
T81.P1	7	bc	31.2	0.94	0.11	66.4	390	Calm	67.5	81.0	29.00	40	55,285	38,898	41,466	64,509	66,815	0.603
T81.P2	7	bc	30.9	0.94	0.11	66.4	388	Calm	66.5	79.5	29.00	36	57,820	35,185	37,043	56,825	59,650	0.619
T101.P1	8	abc	24.8	1.03	0.07	70.2	395	Calm	58.0	75.0	28.99	32	30,194	3,373	2,922	23,470	26,640	0.144
T101.P2	8	abc	25.0	1.02	0.07	69.9	395	Calm	58.0	77.5	28.98	19	23,078	3,328	2,890	23,251	23,006	0.143
T104.P1	8	abc	24.7	1.03	0.07	69.8	380	Calm	58.0	74.0	28.74	15	16,413	2,829	2,263	15,885	16,539	0.178
T104.P2	8	abc	25.0	1.03	0.08	69.6	382	Calm	58.0	76.0	28.74	13	14,430	2,140	1,746	12,273	12,280	0.174
T99.P1	9	d	89.4	0.38	0.48	0.5	305	Calm	59.8	75.5	28.54	3	4,979	2,407	8,947	12,498	12,903	0.193
T99.P2	9	d	89.4	0.38	0.49	0.4	305	Calm	58.8	75.8	28.54	1	5,168	2,455	9,128	12,848	12,455	0.191
T100.P1	9	d	91.9	0.39	0.48	0.4	308	Calm	55.5	71.0	28.96	3	12,577	6,802	24,351	34,586	34,690	0.197
T100.P2	9	d	92.1	0.39	0.48	0.4	305	Calm	55.0	72.0	28.98	26	15,067	6,837	24,743	34,186	34,553	0.200
T76.P1	10	ad	44.6	1.02	0.18	0.6	342	Calm	63.5	75.0	29.04	2	11,026	170	60	41,169	41,700	0.004
T76.P2	10	ad	44.4	1.02	0.18	0.4	342	Calm	63.8	75.0	29.03	9	11,909	145	56	39,353	39,877	0.004
T98.P1	10	ad	44.0	1.01	0.18	0.4	325	Calm	58.0	72.5	28.49	5	1,916	51	12	6,646	6,864	0.008
T98.P2	10	ad	43.4	1.01	0.18	0.4	330	Calm	58.0	74.0	28.49	4	1,436	48	9	5,582	5,362	0.009
T102.P1	11	bd	32.2	1.16	0.11	0.5	338	Calm	58.5	72.0	28.73	1	2,630	648	5,688	9,043	8,797	0.072
T102.P2	11	bd	32.0	1.16	0.11	0.6	338	Calm	57.5	72.0	28.74	1	2,860	762	6,825	10,826	10,591	0.070
T108.P1	11	bd	32.4	1.21	0.11	0.2	345	Calm	60.5	76.0	29.08	18	5,904	1,295	11,921	20,263	20,893	0.064
T108.P2	11	bd	31.9	1.21	0.11	0.2	345	Calm	60.0	77.0	29.08	14	5,290	1,204	11,147	18,852	19,243	0.064
T89.P1	12	abd	24.6	1.25	0.08	0.4	345	Calm	59.5	76.5	28.84	16	8,061	91	73	24,205	30,688	0.004
T89.P2	12	abd	25.0	1.25	0.08	0.4	340	Calm	58.0	76.0	28.54	5	5,670	78	63	20,518	20,824	0.004
T106.P1	12	abd	24.6	1.24	0.07	0.4	340	Calm	59.0	74.5	28.80	7	6,586	65	68	22,031	22,337	0.003
T106.P2	12	abd	24.6	1.24	0.07	0.5	342	Calm	58.5	74.5	28.80	2	4,884	60	61	21,025	21,388	0.003
T77.P1	13	cd	92.2	0.40	0.45	40.1	312	Calm	63.2	75.0	29.04	22	9,644	4,143	15,289	22,342	23,276	0.185
T77.P2	13	cd	91.6	0.40	0.45	40.2	312	Calm	62.5	75.0	29.04	2	8,614	3,911	14,549	20,907	20,859	0.187
T83.P1	13	cd	93.0	0.40	0.44	40.9	312	Calm	65.0	77.5	29.12	7	27,267	14,755	51,099	71,685	71,790	0.206
T83.P2	13	cd	93.4	0.40	0.44	40.9	310	Calm	65.0	78.0	29.12	1	31,608	15,275	52,470	72,480	71,371	0.211

[†]The mass flowmeter analog output had a several tenths of flow bias at 0.0 L/min on the display.
[‡]The particle counter concentrations are for particle diameter sizes ranging from 0.3 to 1.0 µm.

Table B-3.—Pressurizer test data—Continued
(Tests sorted by run number or experimental condition)

Test and period Nos.	Run No.	Test condition	Cab operating parameters					Wind	Wet-bulb temp., °F	Dry-bulb temp., °F	Barometric press., in Hg	Decay time, min	Start †C_1 counts/L	Last 15-min test average			Test average †C_3 counts/L	Cab Pen C_1/C_3
			Q_t ft³/min	$-\Delta p_f$ in w.g.	†Q_L L/min	$+\Delta p_c$ in w.g.	Q_R ft³/min	v_w ft/min						†C_1 counts/L	†C_2 counts/L	†C_3 counts/L		
T86.P1	13	cd	92.8	0.38	39.8	0.49	315	Calm	56.5	73.5	29.22	11	8,467	3,873	14,435	21,633	21,977	0.179
T86.P2	13	cd	92.5	0.38	39.6	0.49	315	Calm	56.5	74.5	29.22	1	8,142	4,254	15,777	22,804	22,264	0.187
T90.P1	14	acd	46.0	0.99	69.0	0.21	340	Calm	60.5	74.5	28.96	21	18,848	611	6,391	57,668	61,294	0.011
T90.P2	14	acd	45.8	0.99	68.7	0.21	338	Calm	59.5	75.0	28.94	17	11,406	445	4,424	40,842	42,917	0.011
T91.P1	14	acd	45.0	0.99	68.3	0.21	332	Calm	63.0	79.5	28.86	2	5,894	306	2,783	26,395	26,748	0.012
T91.P2	14	acd	44.8	0.99	68.1	0.21	330	Calm	61.5	80.0	28.84	4	5,600	291	2,634	25,086	24,960	0.012
T88.P1	15	bcd	34.8	1.13	73.5	0.13	332	Calm	61.0	75.0	28.84	21	4,449	969	8,149	12,759	13,219	0.076
T88.P2	15	bcd	34.7	1.13	73.5	0.13	335	Calm	60.0	75.5	28.84	16	3,867	871	7,302	11,316	11,691	0.077
T97.P1	15	bcd	33.8	1.13	73.7	0.12	335	Calm	57.0	75.5	28.69	14	21,084	4,766	40,289	60,301	62,843	0.079
T97.P2	15	bcd	33.6	1.13	73.5	0.12	332	Calm	56.0	76.0	28.68	1	18,532	5,130	43,501	64,722	62,175	0.079
T82.P1	16	abcd	28.0	1.23	77.0	0.09	330	Calm	63.5	75.0	29.11	2	7,680	511	5,207	35,108	32,679	0.015
T82.P2	16	abcd	27.9	1.24	77.1	0.09	335	Calm	63.0	74.0	29.12	1	9,891	1,338	14,985	89,881	77,690	0.015
T84.P1	16	abcd	28.6	1.25	77.6	0.09	348	Calm	55.8	73.2	29.36	13	1,695	117	836	6,507	6,562	0.018
T84.P2	16	abcd	28.1	1.25	77.4	0.09	350	Calm	55.0	73.8	29.38	1	1,875	131	1,085	8,188	7,758	0.016

†The mass flowmeter analog output had a several tenths of flow bias at 0.0 L/min on the display.
‡The particle counter concentrations are for particle diameter sizes ranging from 0.3 to 1.0 μm.

APPENDIX C.—STEPWISE REGRESSION ANALYSIS OF FILTRATION SYSTEM WITHOUT PRESSURIZER

A stepwise linear regression analysis of the dependent variable (ln *Pen*) with respect to the single factors and two-factor interactions was conducted using SPSS Version 15.0 for Windows. Table C-1 shows the regression model coefficients with their statistical significance, and Table C-2 shows the ANOVA model statistics. The stepwise regression model parameters or coefficients shown in Table C-1 were successively selected by the highest level of significance on cab penetration with no variable removal in the process. The stepwise regression analysis provided a very efficient model with coefficient of multiple determinations (standard and adjusted) above 0.98 for the cab filtration system configured without a pressurizer. Figure C-1 illustrates the goodness of fit of the regression model to the observed response variables. Figures C-2 and C-3 illustrate that the normality and equal variance assumptions were reasonably met by the natural logarithm transformation of cab penetrations for the regression model. This regression model is considered reasonably good, but others could be formulated from these experiments.

Table C-1.—Stepwise regression model without pressurizer

$R^2 = 0.983$, $R^2_{adj} = 0.982$, $n = 148$
Standard error of regression = 0.220, Durbin-Watson statistic = 1.173

Regression model ln *Pen* =	Coefficient	Standard error	*t*-statistic	Significance level
Intercept	−2.598	0.018	−142.145	0.000
Recirculation filter (*D*)	−1.146	0.018	−62.349	0.000
Intake filter efficiency (*A*)	−1.108	0.018	−60.744	0.000
Intake filter efficiency × loading (*AB*)	0.260	0.018	14.203	0.000
Leakage (*C*)	0.230	0.018	12.578	0.000
Intake filter efficiency × leakage (*AC*)	0.201	0.018	10.932	0.000
Intake filter efficiency × recirculation filter (*AD*)	−0.168	0.018	−9.136	0.000
Loading × recirculation filter (*BD*)	−0.162	0.018	−8.766	0.000
Loading × wind (*BE*)	0.050	0.018	2.728	0.007
Loading × leakage (*BC*)	−0.040	0.018	−2.198	0.030

Table C-2.—ANOVA for stepwise regression model without pressurizer

Regression model	Sum of squares	Degrees of freedom	Mean square	*F*-statistic	Significance level
Regression	396.161	9	44.018	906.988	0.000
Residual	6.697	138	0.049	—	—
Total	402.858	147	—	—	—

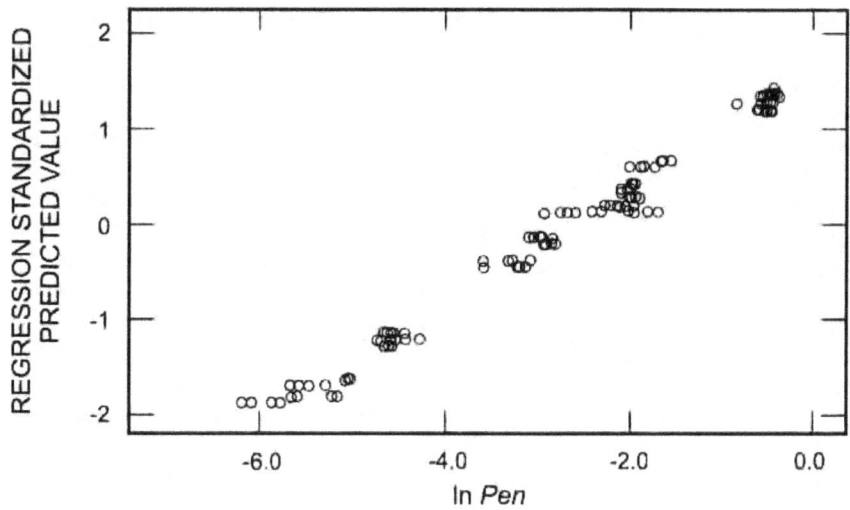

Figure C-1.—Standardized predicted values for regression model without pressurizer.

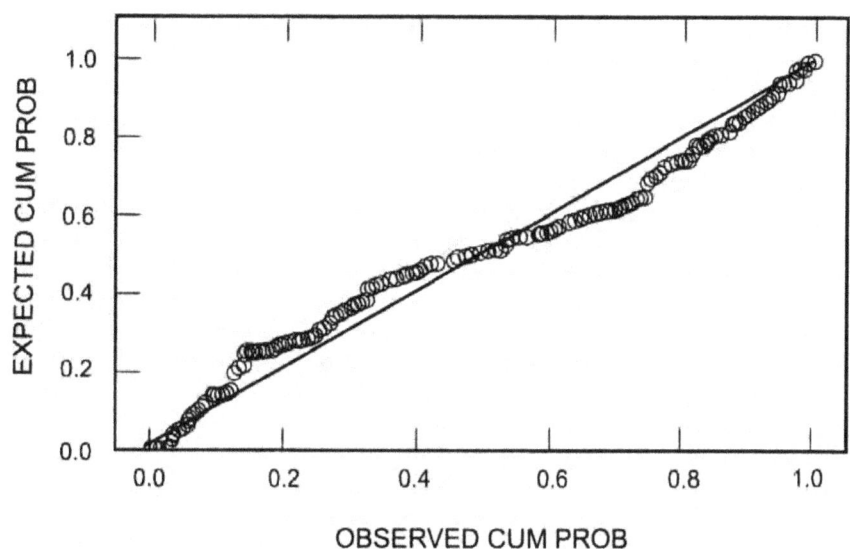

Figure C-2.—Normal probability plot of standardized residuals without pressurizer.

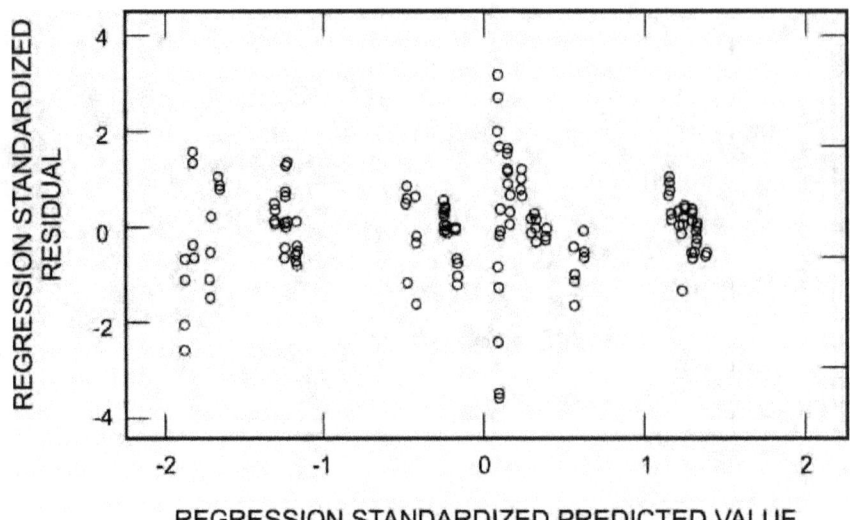

Figure C-3.—Standardized residuals versus standardized predicted values without pressurizer.

APPENDIX D.—STEPWISE REGRESSION ANALYSIS OF FILTRATION SYSTEM WITH AND WITHOUT PRESSURIZER

Regression analysis was conducted with the additional pressurizer test data to statistically examine the effect on cab penetration. This regression analyzed the enclosed cab filtration system data with and without an intake pressurizer fan and no wind. The cab filtration tests without the pressurizer were considered one block of experiments coded with −1. The cab filtration tests with the pressurizer (P) were considered another block of experiments coded with +1. A stepwise linear regression analysis of the dependent variable (ln Pen) with respect to the intake pressurizer blocks, the single factors, and the two-factor interactions within the blocks was conducted using SPSS Version 15.0 for Windows. Table D-1 shows the regression model coefficients with their statistical significance, and Table D-2 shows the ANOVA model statistics. The stepwise regression model parameters or coefficients shown in Table D-1 were successively selected by the highest level of significance on cab penetration with no variable removal in the process. Figure D-1 shows the plot of the standardized predicted values for the regression model. Figure D-2 shows the normal probability plot of the standardized residuals, and Figure D-3 shows the plot of standardized residuals versus standardized predicted values. This regression model is considered reasonably good, but others could be formulated from these experiments.

Table D-1.—Stepwise regression model with and without pressurizer

$R^2 = 0.988$, $R^2_{adj} = 0.976$, $n = 144$
Standard error of regression = 0.260, Durbin-Watson statistic = 1.085

Regression model ln Pen =	Coefficient	Standard error	t-statistic	Significance level
Intercept	−2.538	0.022	−116.028	0.000
Intake filter efficiency (A)	−1.156	0.022	−52.844	0.000
Recirculation filter (D)	−1.075	0.022	−49.057	0.000
Leakage (C)	0.241	0.022	11.011	0.000
Intake filter efficiency × loading (AB)	0.206	0.022	9.416	0.000
Intake filter efficiency × leakage (AC)	0.217	0.022	9.890	0.000
Loading × recirculation filter (BD)	−0.176	0.022	−8.026	0.000
Intake filter efficiency × recirculation filter (AD)	−0.157	0.022	−7.172	0.000
Loading (B)	−0.067	0.022	−3.058	0.003
Pressurizer (P)	0.051	0.022	2.329	0.021

Table D-2.—ANOVA for stepwise regression model with and without pressurizer

Regression model	Sum of squares	Degrees of freedom	Mean square	F-statistic	Significance level
Regression	364.014	9	40.446	597.659	0.000
Residual	9.068	134	0.068	—	—
Total	373.082	143	—	—	—

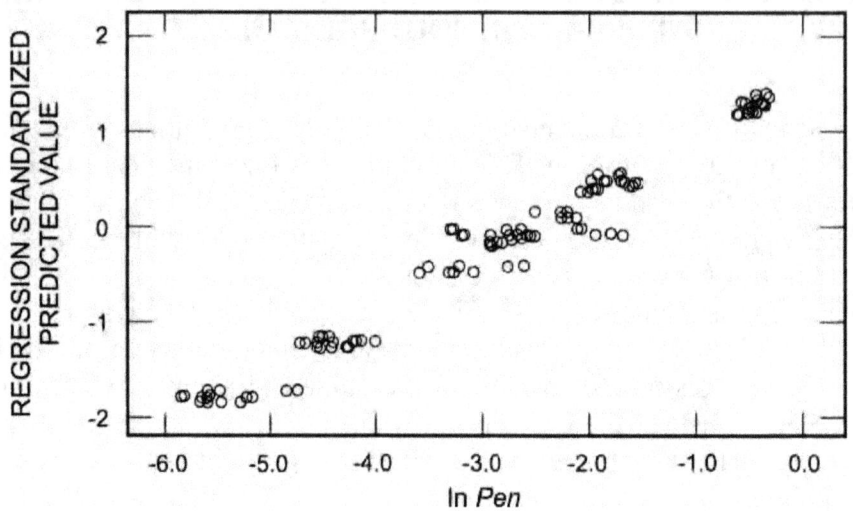

Figure D-1.—Standardized predicted values for regression model with and without pressurizer.

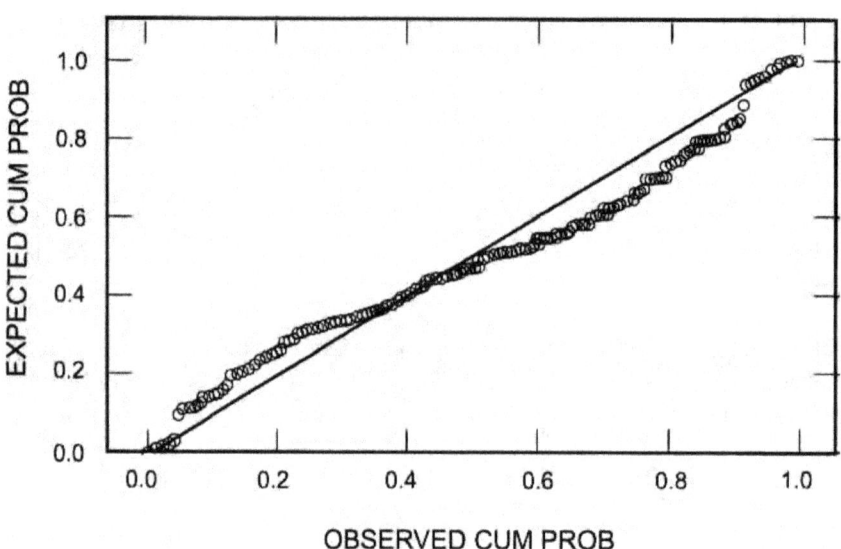

Figure D-2.—Normal probability plot of standardized residuals with and without pressurizer.

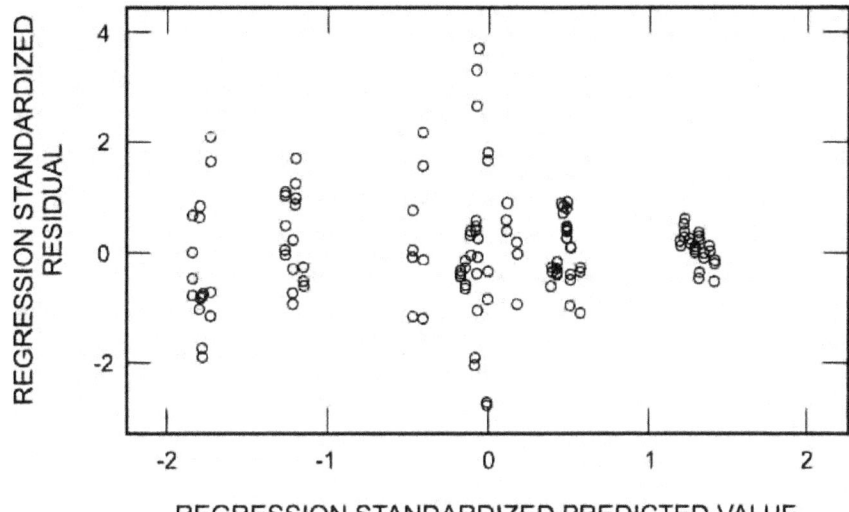

Figure D-3.—Standardized residuals versus standardized predicted values with and without pressurizer.

APPENDIX E.—MATHEMATICAL MODEL FOR CAB FILTRATION SYSTEM

Development of this cab filtration model is based on a time-dependent mass balance model of airborne substances within a control volume. Equation E-1 below is a differential equation describing the mass balance of an airborne substance in a cab filtration system control volume shown in Figure E-1. This is a reformulation of the basic equation for general dilution ventilation [Hartman 1961]. The left-hand part of the equation describes the mass of the contaminant in the control volume. The positive terms in the right-hand part of the equation describe the addition of contaminant mass into the control volume, including intake air leakage, intake filter penetration, and wind infiltration. The negative terms describe the removal of the contaminant mass from the control volume, including intake air dilution and recirculation filter removal.

Mathematical model:

$$V_c\,dx = Q_L C\,dt + Q_F C(1-\eta_I)\,dt + Q_w C\,dt - Q_I x\,dt - Q_R x \eta_R\,dt \qquad (E\text{-}1)$$

Model assumptions:

(1) Outside contaminant concentration is constant.
(2) Contaminant leakage into the filtration system is proportional to the air quantity leakage around the filter.
(3) Wind penetration into the cab occurs when the wind velocity (v_w) exceeds the opposing cab exit air velocity (v_I).

where
- V_c = cab volume,
- x = inside cab contaminant concentration,
- Q_F = filtered intake air quantity,
- η_I = intake filter efficiency, fractional,
- Q_L = air leakage quantity around the intake filter,
- Q_I = intake air quantity into the cab,
- l = portion of intake air leakage, or Q_L/Q_I,
- Q_R = recirculation filter airflow,
- η_R = recirculation filter efficiency, fractional,
- C = outside cab contaminant concentration,
- Q_w = wind quantity infiltration into the cab,
- t = time,
- v_I = cab intake air exit velocity,

and
- v_w = wind velocity.

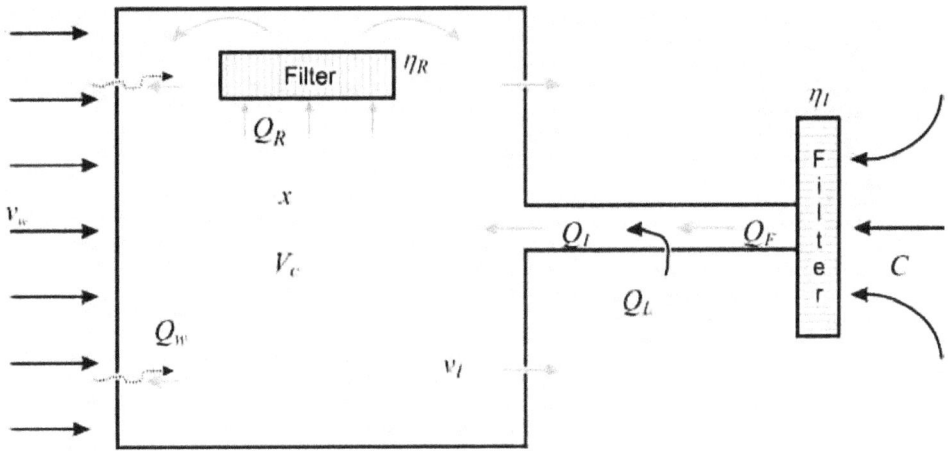

Figure E-1.—Schematic of basic cab filtration system.

Since: $Q_I = Q_L + Q_F$ and $Q_L = Q_I l$; $Q_F = Q_I (1-l)$

$$V_c dx = Q_I l C dt + Q_I(1-l)(1-\eta_I)C dt + Q_w C dt - (Q_I + Q_R \eta_R) x dt \qquad (E\text{-}2)$$

$$V_c dx = Q_I C(1-\eta_I + l\eta_I) dt + Q_w C dt - (Q_I + Q_R \eta_R) x dt \qquad (E\text{-}3)$$

$$\int_{x_o}^{x} \frac{dx}{Q_I C(1-\eta_I + l\eta_I) + Q_w C - (Q_I + Q_R \eta_R) x} = \frac{1}{V_c} \int_{t_1}^{t_2} dt \qquad (E\text{-}4)$$

Let: $u = Q_I C(1-\eta_I + l\eta_I) + Q_w C - (Q_I + Q_R \eta_R) x$ and $du = -(Q_I + Q_R \eta_R) dx$

$$\frac{1}{-(Q_I + Q_R \eta_R)} \int_{u_o}^{u} \frac{du}{u} = \frac{1}{V_c} \int_{t_1}^{t_2} dt \qquad (E\text{-}5)$$

Integrate and rearrange:

$$\ln|u| - \ln|u_o| = \frac{-(Q_I + Q_R \eta_R)(t_2 - t_1)}{V_c} \qquad (E\text{-}6)$$

$$\ln \frac{u}{u_o} = \frac{-(Q_I + Q_R \eta_R) \Delta t}{V_c} \qquad (E\text{-}7)$$

Substitute for u:

$$\ln\frac{Q_I C(1-\eta_I+l\eta_I)+Q_w C-(Q_I+Q_R\eta_R)x}{Q_I C(1-\eta_I+l\eta_I)+Q_w C-(Q_I+Q_R\eta_R)x_o}=\frac{-(Q_I+Q_R\eta_R)\Delta t}{V_c} \quad \text{(E-8)}$$

$$\frac{Q_I C(1-\eta_I+l\eta_I)+Q_w C-(Q_I+Q_R\eta_R)x}{Q_I C(1-\eta_I+l\eta_I)+Q_w C-(Q_I+Q_R\eta_R)x_o}=e^{\left(\frac{-(Q_I+Q_R\eta_R)\Delta t}{V_c}\right)} \quad \text{(E-9)}$$

The steady-state solution as $\Delta t \to \infty$; $e^{-\infty} \to 0$

$$Q_I C(1-\eta_I+l\eta_I)+Q_w C-(Q_I+Q_R\eta_R)x=0 \quad \text{(E-10)}$$

$$x=\frac{Q_I C(1-\eta_I+l\eta_I)+Q_w C}{Q_I+Q_R\eta_R} \quad \text{(E-11)}$$

$$Pen=\frac{x}{C}=\frac{Q_I(1-\eta_I+l\eta_I)+Q_w}{Q_I+Q_R\eta_R} \quad \text{(E-12)}$$

Rearrange into other useful forms:

$$Pen=\frac{1-\eta_I+l\eta_I+\dfrac{Q_w}{Q_I}}{1+\dfrac{Q_R}{Q_I}\eta_R} \quad \text{(E-13)}$$

$$Pen=\frac{1-\eta_I+\dfrac{Q_L}{Q_I}\eta_I+\dfrac{Q_w}{Q_I}}{1+\dfrac{Q_R}{Q_I}\eta_R} \quad \text{(E-14)}$$

The air quantity leakage around the filter (Q_L) and the wind quantity infiltration into the cab (Q_w) in the above equations may be estimated by applying orifice flow equations derived from Bernoulli's principle [Streeter and Wylie 1979; Heitbrink et al. 2000]. The orifice flow relationship for air at atmospheric and turbulent flow conditions is shown in Equation E-15 [Streeter and Wylie 1979], assuming the air is incompressible with a Reynolds number ≥ 4000. A particularly developed wind infiltration relationship into cabs when the wind velocity pressure exceeds cab static pressure is also shown in Equation E-16 [Heitbrink et al. 2000].

$$v_o = \frac{Q_o}{A_o C_d} = \sqrt{2\frac{\Delta p_o}{\rho_{air}}} \; ; \quad \text{orifice flow from high to low pressure} \tag{E-15}$$

$$v_o = \frac{Q_o}{A_o} = 0.61\sqrt{2\frac{(0.5\rho_{air} v_w^2 - p_c)}{\rho_{air}}} \; ; \quad \text{wind penetration when } 0.5\rho_{air} v_w^2 > p_c \tag{E-16}$$

where
- v_o = fluid velocity through an orifice,
- Q_o = airflow quantity through an orifice,
- C_d = orifice discharge coefficient,
- A_o = area of orifice,
- Δp_o = air pressure differential across orifice,
- ρ_{air} = air density,
- v_w = wind velocity,

and
- p_c = cab pressure.

www.ingramcontent.com/pod-product-compliance
Lightning Source LLC
Chambersburg PA
CBHW081909170526
45167CB00007B/3211